Hydrogels and Nanogels - Applications in Medicine

Edited by Chukwuebuka Umeyor, Emmanuel Uronnachi and Pratik Kakade

Published in London, United Kingdom

Hydrogels and Nanogels - Applications in Medicine
http://dx.doi.org/10.5772/intechopen.1000374
Edited by Chukwuebuka Umeyor, Emmanuel Uronnachi and Pratik Kakade

Contributors
Hany F. Nour, Ahmed M. Salama, Badria H.A. Al-Dhuwayin, Amal F. Seliem, Haddad Antonia-Nancy, Michiaki Matsumoto, Yoshiro Tahara, Asif Mahmood, Hira Ijaz, Rai Muhammad Sarfraz, Nadiah Zafar, Muhammad Zaman, Mariya Azam, Ruby Varghese, Yogesh Bharat Dalvi, P. Lochana, S. Achinthya, Bhagyashri Omprakash Somani, Preetha Karnaver, Nebu George Thomas, S. Rupesh, Nibu Varghese, Jayachandran V.P., Mohammed Hussain AL-Mayahy, Hiba Imad Hameed, Mansurat Oluwatoyin Shoge, Chukwuebuka Umeyor, Emmanuel Uronnachi, Abhit Nayak, Tochukwu Okeke, Purav Shah, Vandana Patravale, Maryam Adenike Salaudeen

First published in London, United Kingdom, 2024 by IntechOpen
IntechOpen is the global imprint of INTECHOPEN LIMITED, registered in England and Wales, registration number: 11086078, 5 Princes Gate Court, London, SW7 2QJ, United Kingdom

British Library Cataloguing-in-Publication Data
A catalogue record for this book is available from the British Library

Additional hard and PDF copies can be obtained from orders@intechopen.com

Hydrogels and Nanogels - Applications in Medicine
Edited by Chukwuebuka Umeyor, Emmanuel Uronnachi and Pratik Kakade
p. cm.
Print ISBN 978-1-83769-793-9
Online ISBN 978-1-83769-792-2
eBook (PDF) ISBN 978-1-83769-794-6

For EU product safety concerns:
IN TECH d.o.o., Prolaz Marije Krucifikse Kozulić 3, 51000 Rijeka, Croatia,
info@intechopen.com or visit our website at intechopen.com.

Meet the editors

Chukwuebuka E. Umeyor is a Senior Research Lecturer in the Department of Pharmaceutics and Pharmaceutical Technology, Faculty of Pharmaceutical Sciences, Nnamdi Azikiwe University, Awka, Nigeria. He holds a BPharm, MPharm, and Ph.D. in Physical Pharmaceutics and Drug Delivery from the University of Nigeria, Nsukka, Enugu State, Nigeria. He is a recipient of an African-German Network of Excellence in Science (AGNES) Junior Researcher Grant, an International Foundation for Science (IFS) grant, and a Department of Biotechnology-The World Academy of Sciences (DBT-TWAS) Postdoctoral Research Fellowship at the Department of Pharmaceutical Sciences and Technology, Institute of Chemical Technology, Mumbai, India. He has several publications in reputable peer-reviewed local and international journals to his credit. His research interests include lipid- and polymeric-based drug and gene delivery systems, tissue engineering, nanovaccines and adjuvants, nanodiagnostics, brain drug delivery, drug targeting, biomaterials, and more.

Dr. Emmanuel Maduabuchi Uronnachi holds a bachelor's degree in Pharmacy and a master's degree and Ph.D. in Physical Pharmaceutics (Formulation Science) from the University of Nigeria, Nsukka, Enugu State, Nigeria. He holds a Postgraduate Diploma in Drug Delivery from the University of Ibadan, Nigeria. He is a reader at the Department of Pharmaceutics and Pharmaceutical Technology, Faculty of Pharmaceutical Sciences, Nnamdi Azikiwe University, Awka, Nigeria. His research interests include nanomedicine, formulation science and drug delivery, with an interest in improving the solubility and bioavailability of poorly soluble drugs. He has several publications in reputable peer-reviewed local and international journals to his credit. Dr. Uronnachi received the Tertiary Education Trust Fund (TETFUND) Nigeria Institutional Based Research grant and a postdoctoral research grant from Prince of Songkla University, Thailand.

Dr. Pratik Kakade received his Ph.D. from the Institute of Chemical Technology, Mumbai, India. He received India's topmost and most prestigious prime minister's doctoral fellowship for his Ph.D. work. He is also the recipient of two national and two international awards for poster and oral presentations. He completed an industrial project on generic, solid, oral dosage forms and has worked on many other industry projects. He holds an MBA in Marketing. Dr. Kakade has more than ten publications and one patent application to his credit.

Contents

Preface

Literature is replete with numerous benefits derived from rational drug administration and usage. At any given time, the goal of any clinical therapeutic intervention is to ensure that the proper dose of the drug is administered to achieve the required maximum concentration of drug molecules at the required site of activity to achieve the desired therapeutic response. Sometimes, this vital objective, which improves the condition of the recipient of the intervention, boosts the confidence of the clinician, and provides the formulator with the feedback needed to sustain the development and formulation of quality medicines for clinical applications, is difficult to realize using the traditional dosage forms owing to factors such as administration of inappropriate dosage form and dose, low drug efficacy, poor drug bioavailability, and poor treatment compliance by the patient, among others. Consequently, there is a constant search for innovative methods to enhance drug delivery for better therapeutic outcomes and circumvent the drawbacks occasioned by the constant use of classic drug delivery strategies.

Whereas the concept of nanotechnology opened novel frontiers for effective delivery of bio-actives to achieve improved biomedical objectives, the use of biomaterials to manufacture biomimetic delivery systems, such as hydrogels and nanogels, revolutionized drug administration because of the numerous desirable possibilities, including enhanced deposition of entrapped drug and prolonged contact time with the site of activity. Hydrogels and nanogels are biocompatible and biodegradable nanomaterials with pliable properties for the delivery of effective concentrations of drugs, vaccines, genes, proteins, and more to achieve desirable therapeutic outcomes. This book, *Hydrogels and Nanogels – Applications in Medicine*, discusses current strategies for synthesis and functionalization of hydrogels and nanogels, provides new insights on the mechanisms of drug release from hydrogels and nanogels, presents techniques to enhance nanostructure-based cellular and tissue interactions and targeting, highlights advanced applications of hydrogels and nanogels in medicine, discusses challenges encountered in the fabrication of hydrogels and nanogels, and proposes future interventions. It presents is a valuable resource for those who work in formulation science on the development and formulation of hydrogels and nanogels for effective and efficient drug delivery, which is expected to result in increased scaling and translation of hydrogel and nanogel delivery systems to commercialized dosage forms and products available for clinical use.

I acknowledge Publishing Process Manager Tea Jelaca at IntechOpen for guiding me through the editorial process. I also thank Professor Vandana Patravale of the Department of Pharmaceutical Sciences and Technology, Institute of Chemical Technology, Mumbai, India, for providing the conducive environment and necessary resources to make this project a reality. I also acknowledge the TWAS-DBT-UNESCO postdoctoral fellowship program for providing the fellowship and travel opportunity to the Institute of Chemical Technology, Mumbai, India.

Chukwuebuka Umeyor
Faculty of Pharmaceutical Sciences,
Department of Pharmaceutics and Pharmaceutical Technology,
Nnamdi Azikiwe University,
Awka, Nigeria

Department of Pharmaceutical Sciences and Technology,
Institute of Chemical Technology,
Mumbai, India

Emmanuel Uronnachi
Faculty of Pharmaceutical Sciences,
Department of Pharmaceutics and Pharmaceutical Technology,
Nnamdi Azikiwe University,
Awka, Nigeria

Drug Delivery System Excellence Centre,
Faculty of Pharmaceutical Sciences,
Prince of Songkhla University,
Hat Yai, Thailand

Pratik Kakade
Department of Pharmaceutical Sciences and Technology,
Institute of Chemical Technology,
Mumbai, India

Chapter 1

Low-Molecular-Weight Hydrogels: Synthetic Methodologies, Gelation Mechanisms, and Biomedical Applications

Hany F. Nour, Ahmed M. Salama, Badria H.A. Al-Dhuwayin and Amal F. Seliem

Abstract

Low-molecular-weight hydrogels (LMWHs) have garnered widespread focus as versatile soft materials owing to their distinctive characteristics and potential applications. LMWHs are synthesized from small molecules that, upon assembly, form entangled aggregates *via* different types of noncovalent interactions, such as hydrogen bonding, van der Waals forces, or π-π stacking interactions. LMWHs are characterized by their unique ability to mimic biological systems by effectively absorbing and retaining large quantities of water. Despite their poor mechanical properties, LMWHs are widely used in various medical applications due to their easy preparation, biocompatibility, and low toxicity. Smart LMWHs demonstrate responsiveness to external stimuli, such as light, temperature, enzymes, or pH, rendering them ideally adapted for various controlled drug delivery applications. LMWHs have been extensively employed in different biomedical applications, including drug delivery, tissue engineering and cell culture, wound healing, and biofabrication. In this chapter, we aim to explore the potential of LMWHs as drug-delivery vehicles for a range of medications, focusing on the different synthetic strategies, gelation processes, and drug-loading and releasing mechanisms.

Keywords: low-molecular-weight hydrogels, smart hydrogels, drug delivery, biomedical applications, stimuli-responsive, gelation mechanisms, noncovalent interactions

1. Introduction

Low-molecular-weight hydrogels (LMWHs) have gained significant attention across various disciplines as an important class of soft materials. These gels are composed of small organic molecules that can self-assemble in water, forming 3D network structures. The self-assembly of LMWHs into entangled networks is typically driven by various noncovalent interactions, such as hydrogen bonding, π-π stacking, or van der Waals forces. The noncovalent interactions between the molecules result in the formation of fibrous structures that become entangled and form hydrogels. LMWHs are known for their high water retention capability, which imparts them with a resemblance to biological tissues. LMWHs have been widely studied in various biomedical research areas due to their biocompatibility, tailored design, cost-effectiveness, and ease of preparation. LMWHs show great potential as a drug delivery platform, as they can encapsulate drugs and release them in a slow and sustained manner. For instance, LMWHs have been used to deliver vascular endothelial growth factor small interfering RNA (VEGF-siRNA) into human cells [1]. LMWHs derived from vitamin B have been utilized in the delivery of siRNA into cancer cell lines with high selectivity [2]. LMWHs have also been exploited for the delivery of proteins [3, 4].

In addition to their use in drug delivery, LMWHs have exhibited significant potential in several medicinal applications. LMWHs with inherent antibacterial properties have been widely explored [5–7]. Adams *et al.* reported the gelation of a naphthalene-based dipeptide, which was triggered by the oxidation of dopamine, with the generation of reactive oxygen species (ROS). The hydrogel thus formed demonstrated potent antibacterial activity against *S. aureus* [8]. The naturally occurring glycyrrhizic acid has been used as an injectable hydrogelator with intrinsic antibacterial activity against *S. aureus* [9]. Kumar *et al.* described the gelation of a peptide LMWH with antibacterial potential against *S. aureus* [10]. LMWHs that incorporate *L*-proline or *D*-phenylalanine along with thiophene were reported to show potential suppression of MRSA [11]. Diimidazolium-based LMWHs have been shown to inhibit different pathogens, such as *K. rhizophila*, *B. subtilis*, and *E. coli* [12]. LMWHs have been evaluated as versatile materials for wound healing applications. LMWHs, when applied to wounds, can function as a physical barrier that prevents fluid loss and offers a moist environment that promotes recovery. An illustration of this is a hydrogel prepared from cytidine, boric acid, and silver nitrate, which has been utilized for wound healing. It has been shown to have antibacterial effects against *S. aureus*, *E. coli*, and *P. aeruginosa* pathogens and has also been found to promote wound healing in mice [13, 14]. LMWHs have shown remarkable success in the development of potent anticancer therapeutics [15]. Marlow *et al.* described the synthesis of LMWHs incorporated with 5-fluoro-benzothiazole, which exhibited high inhibition of breast and ovarian cancers [16]. Amphiphilic phenylalanine gelators, when mixed with Au(III) ions in water, have been shown to generate ROS that inhibit mammalian cancer cells [17]. The use of LMWHs in tissue engineering and 3D cell cultures has undergone remarkable progress [18]. Lampe *et al.* demonstrated the synthesis of pentapeptide hydrogels as the supportive matrix for the cell culture of oligodendrocyte progenitor cells [19]. The oligodendrocyte progenitor cells encapsulated in the hydrogel matrix exhibited proliferation and extension, indicating the potential utility of LMWHs in tissue engineering applications. LMWHs have also been employed in biofabrication and cell adhesion applications [20–22].

This chapter aimed to provide a comprehensive understanding of the design, synthesis, and applications of LMWHs in drug delivery. We discussed the various synthetic methods utilized for the fabrication of LMWHs and highlighted the challenges of using LMWHs as drug carriers along with possible solutions. In addition, we reviewed the different drug cargos that have been utilized in drug delivery purposes incorporating LMWHs. These drugs included anticancer, anti-inflammatory, antibacterial agents, and vitamins, to name a few. The versatility and potential of LMWHs as drug-delivery vehicles in various therapeutic areas were emphasized. Furthermore, we provided insights into the mechanism of drug release from LMWHs. We also discussed the influence of the chemical structure of LMWHs on drug delivery efficiency, highlighting in particular, the importance of their functionalization.

2. Exploring the gelation process of LMWHs

LMWHs can form by self-assembly of small molecular weight molecules into 3D network structures capable of immobilizing water. The gelation process can be optimized by testing different parameters and observing their influences on gelation. These parameters include, but are not limited to, the concentration of the gelator, the temperature of gelation, the pH of the solution, and the addition of an external agent to promote gelation. Optimizing these parameters can aid in fine-tuning the conditions required for the organic molecules to dissolve freely in water upon heating and form a clear solution. Upon cooling, the solution can turn into a gel, which can be studied to explore the factors that promote and reinforce the gelation. The inclusion of functional motifs, or groups capable of establishing hydrogen bonding or π-π stacking, into the structural backbone of the gelator is also crucial for promoting the gelation process [23]. Therefore, by testing and optimizing these parameters, the gelation process of LMWHs can be improved and tailored for specific applications.

3. LMWHs in drug delivery: Opportunities and challenges

LMWHs possess several advantages that make them excellent candidates for various drug delivery applications, such as biocompatibility, ease of preparation, structural customization, stimuli-responsiveness, versatility, and controlled drug release. When drugs are loaded into the hydrogel matrix, they gain extra stability against degradation, oxidation, and other adverse effects, leading to a longer shelf life. Additionally, controlled drug release is a crucial benefit of using drug-loaded LMWHs, as it reduces dosing frequency. LMWHs can be tailored to deliver drugs to specific areas, which is especially useful in the case of anticancer medications, as it helps avoid harming healthy cells. Injecting anti-inflammatory and analgesic drugs into sites of injury or inflammation can also provide prolonged pain relief. Moreover, the high water content of LMWHs makes them ideal for encapsulating various drugs. The synthesis of LMWHs is often straightforward and cost-effective when compared to other drug delivery systems.

Despite their intriguing properties, using LMWHs for drug delivery may have some critical limitations. For example, some LMWHs may have weak mechanical strength, a limited capacity for drug loading, and insufficient drug release, which could influence their therapeutic effectiveness. In addition, some LMWHs may not be suitable for encapsulating drugs with poor solubility or achieving targeted drug delivery. Furthermore, certain LMWHs may have low stability for in *vivo* studies, and their efficiency may be affected by changes in pH or temperature, making them less reliable.

To improve the mechanical strength of hydrogels co-assembly can be utilized. For instance, acylhydrazone-based hydrogels incorporating bipyridinium units have been found to form highly stable hydrogels through donor-acceptor charge transfer interactions with electron-rich naphthyl moieties [24, 25]. The addition of metal ions can also enhance the mechanical properties and thermal stability of LMWHs [26].

The design of LMWHs can greatly impact their loading capacity, pH and temperature stability, targeted delivery ability, *in vivo* performance, and drug solubility. Therefore, it is important to customize the design of LMWHs to achieve the desired properties for specific applications.

Drug loading into the hydrogel matrix can be accomplished using physical entrapment or chemical modification of the structure of the hydrogelator, with the former relying on noncovalent interactions and the latter utilizing covalent binding. Nevertheless, both methods have notable drawbacks. Certain drugs may not establish strong enough interactions with the hydrogelator, which can result in rapid drug release, while others may form overly strong interactions, leading to very slow drug release. Chemical modification of the gelators to integrate drugs into their structural framework can lead to a decrease in the drug's effectiveness after its release from the hydrogel structure. For example, the incorporation of ciprofloxacin into the structure of a peptide gelator resulted in a reduction of its antibacterial activity [27]. Therefore, it is highly important to further investigate the process of linking drugs into the backbone structure of the gelators as a drug delivery approach. Additionally, it is crucial to ensure controlled and sustained drug release, especially for physically entrapped drugs, in order to maintain the drugs' efficacy and optimize therapeutic outcomes.

Another important challenge facing drug delivery using LMWHs is the limited drug-loading capacity. This obstacle can be overcome by modifying the hydrogel structure, such as by incorporating functional motifs or groups that reinforce the drug's affinity for the hydrogelator. Moreover, optimizing drug concentration can also impact the drug-loading capacity. The drug loading can also be improved by increasing the porosity of the gelator. In addition, considering the pre-loading or post-loading of drugs into the hydrogel matrix is a critical factor in achieving a high loading capacity. These strategies are essential to developing effective drug delivery systems with higher drug loading capacities.

4. Applications of LMWHs in drug delivery

Hydrogel **3** was prepared from the condensation of tricarbohydrazide **1** with 3,4-disubstituted-benzaldehyde **2**, with the formation of acylhydrazone linkages as shown in **Figure 1** [28]. Augmented hydrogen bonding of the acylhydrazone functional motifs prompted gelation of derivative **3** [29]. *L*-Histidine methyl ester catalyzed the reaction as a biocompatible organocatalyst at a neutral pH of 7. The anticancer drug doxorubicin (DOX) was loaded into the hydrogel matrix either covalently or noncovalently based on the sequence of the reactants mixing. The drug release was accomplished in a buffer solution and peaked at pH 5 due to the hydrolysis of acylhydrazone bonds.

The biocompatibility of the gel was evaluated *in vitro* on the breast cancer cell line (MCF-7) using the 3-(4,5-dimethyl-2-thiazolyl)-2,5-diphenyl-2-tetrazolium bromide (MTT) assay. The MCF-7 cancer cell line treated with the hydrogel-DOX matrix exhibited a 50–70% reduction in cell viability over time, indicating a slow release of DOX.

Banerji *et al.* reported the synthesis of a 2,5-diketopiperazine peptide 7, which gelates at a physiological pH of 7.46 at 37°C and self-assembles in water into nanofibrillar structures [30]. Cyclic peptides have a variety of intriguing features, such as biocompatibility, resistance to degradation, and rigidity. Such interesting properties enrich their biomedical applications. Hydrogelator 7 was prepared by reacting

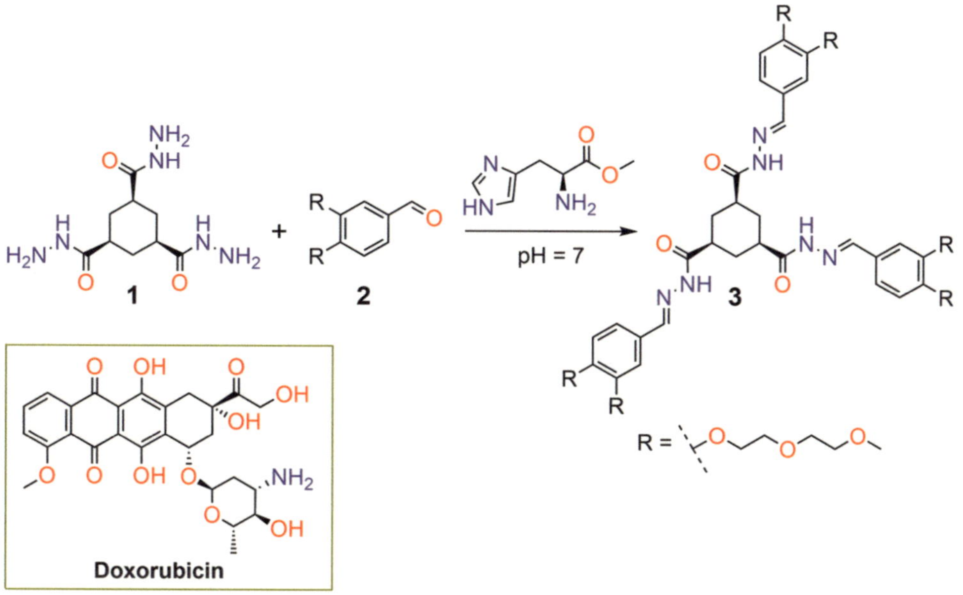

Figure 1.
Synthetic pathway of hydrogelator **3** *and the chemical structure of the drug molecule [28, 29].*

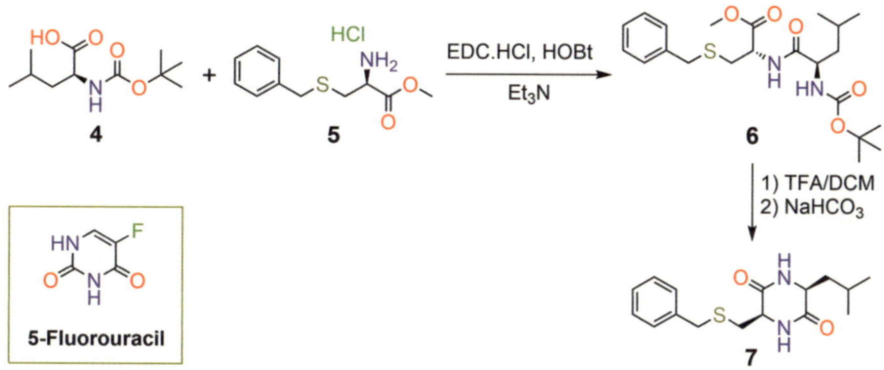

Figure 2.
Synthetic pathway of hydrogelators **7** *and the chemical structure of the drug molecule [30].*

N-(*tert*-butoxycarbonyl)-*L*-leucine **4** with *S*-benzyl-*L*-cysteine methyl ester **5** in the presence of 1-ethyl-3-(3-dimethylaminopropyl)carbodiimide (EDC) and hydroxybenzotriazole (HOBt) to give compound **6**, as illustrated in **Figure 2**. Then, derivative **6** was hydrolyzed using trifluoroacetic acid (TFA) to afford the cyclized product 7. Compound 7 forms a stable hydrogel at a critical gelation concentration (CGC) of 0.05% *w*/*v*. The backbone structure of compound 7 incorporates two amide functionalities, a thiobenzyl group, and an isobutyl chain, which promotes hydrogelation *via* augmented intermolecular hydrogen bonding, π-π stacking, and hydrophobic interactions. Hydrogelator 7 was employed as a drug delivery carrier for the anticancer drug 5-fluorouracil. The encapsulation efficiency of the hydrogelators 7 was reported to be 94.46 ± 1.04%. Drug release occurred by treating the drug-loaded hydrogel with a phosphate-buffered saline solution at a pH of 7.46, which resulted in

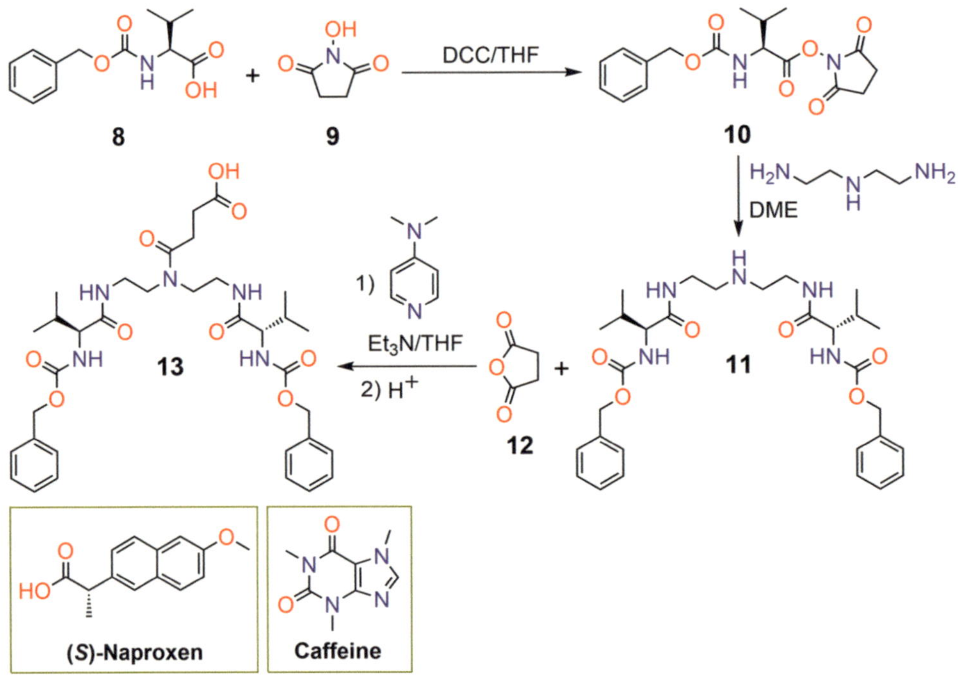

Figure 3.
Synthetic pathway of hydrogel **13** *and the chemical structures of the drug molecules [31, 32].*

90% release. Interestingly, the drug-loaded hydrogel showed higher anticancer activity against the HCT116 human colon cancer cell with an IC_{50} of 6.7 ± 1.2 μM when compared to 5-fluorouracil (IC_{50} = 32.23 ± 3.4 μM), possibly due to the slow release of the drug [30].

Luis *et al.* described the synthesis of pseudopeptide hydrogelator **13** with a C_2-symmetrical structure (**Figure 3**). Gelator **13** functions as a drug delivery carrier for (*S*)-naproxen and caffeine. The reaction of protected *N*-Cbz-*L*-valine **8** with *N*-hydroxysuccinimide **9** in the presence of the dicyclohexylcarbodiimide (DCC) coupling agent yields compound **10** [31]. Derivative **10** reacted with a diamine in dimethoxyethane (DME) to afford **11**.

Compound **11** reacted with succinic anhydride in an alkaline medium to give the carboxylic-functionalized derivative **13** [32]. Hydrogelation of compound **13** was conducted in an acidic condition with a reported CGC of 1 mg/mL. Gelator **13** showed dual thermoresponsive and pH-responsive sol-to-gel transitions. The drug-loaded hydrogels were evaluated for transdermal drug release in animals, which revealed slow and sustained pH-dependent drug-releasing potential.

Ionic hydrogels are a readily available and cost-effective class of materials that have been widely explored as significant drug delivery systems. Cetylpyridinium salicylate ionogel **16** was prepared from the reaction of cetylpyridinium chloride **14** with sodium salicylate **15** in a mixture of H_2O and acetone (**Figure 4**) [33]. Cetylpyridinium salicylate **16** gelates in water after standing at 25°C for 12 hours in the form of long fibers [34]. The CGC of hydrogel **16** was reported to be 4.7% *w*/*v*. The ability of gelator **16** to assemble into fibrous aggregates is mainly driven by non-covalent hydrogen bonding and hydrophobic interactions. Hydrogelator **16** exhibited high potential to encapsulate the anticancer drug imatinib mesylate. The drug-loaded

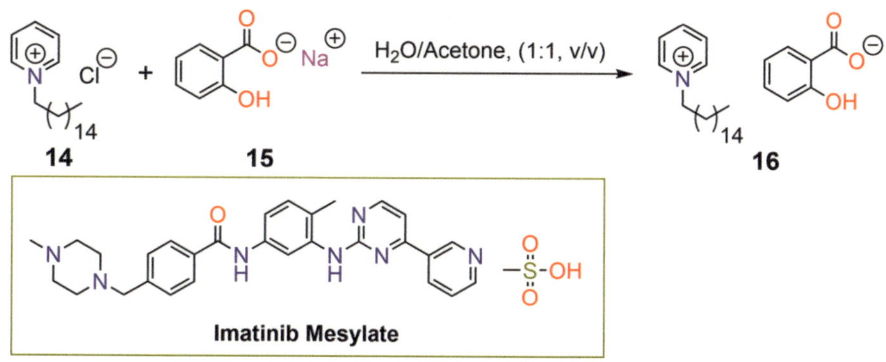

Figure 4.
Synthetic pathway of ionogel **16** *and the chemical structure of the drug molecule [33, 34].*

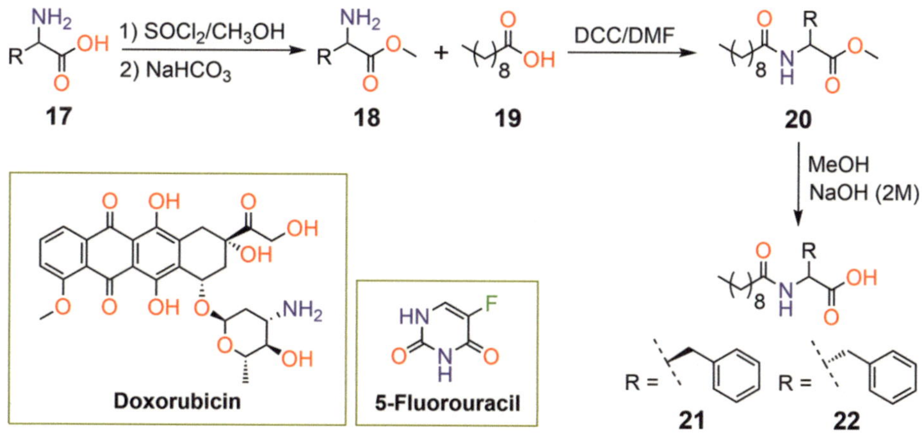

Figure 5.
Synthetic pathway of hydrogels **21** *and* **22** *and the structures of the drug molecules [35].*

ionogel exhibited 53.17, 88.30, and 94.17% releasing at pH 10, 7.4, and 5, respectively, at a typical temperature of 37°C.

DuttKonar *et al.* reported the preparation of chiral hydrogelators **21** and **22**, which serve as drug delivery carriers for the anticancer drugs 5-fluorouracil and DOX. As shown in **Figure 5**, phenylalanine **17** is esterified with thionyl chloride in methanol to give compound **18**; this product, along with decanoic acid **19** and the coupling agent DCC, reacted to yield **20**. The basic hydrolysis of **20** afforded derivatives **21** and **22**.

Compounds **21** and **22** underwent self-assembly, resulting in the formation of entangled fibrous aggregates facilitated by hydrogen bonding, hydrophobic, and π-π stacking interactions [35]. The CGCs of hydrogels **21** and **22** were found to be 0.01% *w/v*, classifying them as supergelators. The hydrogelators exhibited great biocompatibility when incubated with cancer cell lines, as determined by the MTT assay, and showed high stability against proteolytic degradation upon incubation with the proteolytic enzyme, proteinase K. Hydrogel nanoparticles of **21** and **22** were utilized as vehicles for drug delivery of 5-fluorouracil and DOX. Following a 45-hour duration, the release percentages of 5-fluorouracil were determined to be 43.67% and 41.23% for **21** and **22**, respectively. In the instance of DOX, the release potential of nanoparticle hydrogelators **21** and **22** after 80 hours was 50.57% and 48.64%, respectively.

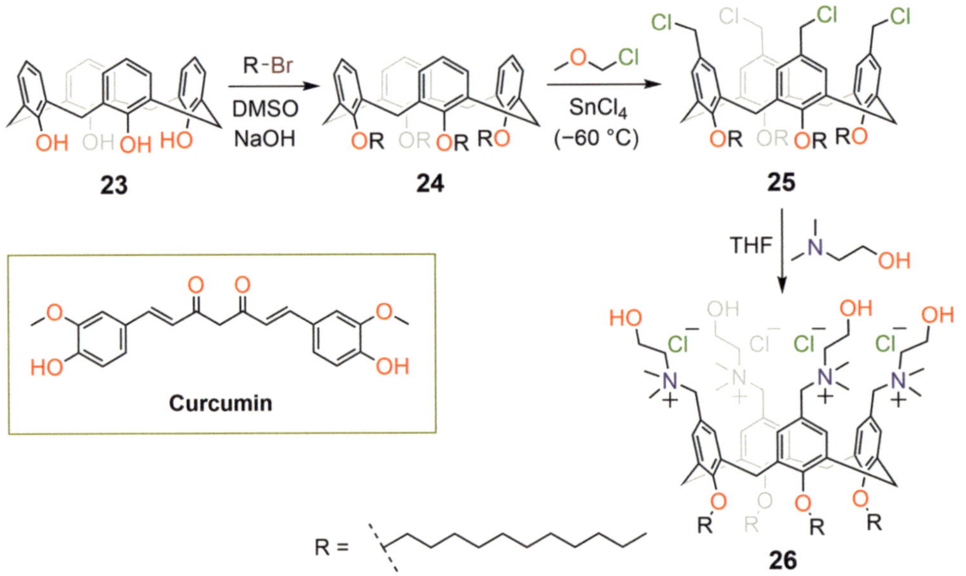

Figure 6.
Synthetic pathway of functionalized calixarene derivative **26** *and the chemical structure of curcumin [36, 37].*

Functionalized calixarene **26** was prepared following the synthetic strategy illustrated in **Figure 6** [36]. Alkylation of calixarene **23** with 1-bromododecane yields the corresponding tetraalkoxy-calixarene **24**.

The tetrachloromethyl-*O*-dodecyl calix[4]arene **25** is formed by the reaction of **24** with chloro(methoxy)methane in the presence of catalytic $SnCl_4$ at −60°C. The reaction of compound **25** with *N*,*N*-dimethylaminoethanol gives derivative **26**. Consoli *et al.* reported the gelation of the polyphenolic curcumin with the cationic choline-calix[4]arene **26**, which has several medicinal applications [37]. Alkoxy chains are attached to the lower rim of calixarene, while the upper rim is functionalized with choline moieties. The calixarene-curcumin hydrogel forms by simple mixing of **26** with curcumin in a phosphate-buffered saline solution at physiological pH of 7.4. The incorporation of curcumin into the gel matrix has resulted in a remarkable enhancement of its solubility, as well as a considerable improvement in its stability against photodegradation. The calixarene-curcumin hydrogel is assembled in the form of nanosphical aggregates. Hydrogelation was assumed to occur *via* hydrogen bonding and ion-dipole interactions, forming 3D network structures. The curcumin showed a slow-releasing potential of 0.30 ± 0.02% per hour from the hydrogel matrix.

Gamble *et al.* described the synthesis of a diphenylalanine-based hydrogel **32** that contains fluorinated-benzyl azide and carbamate moieties [38]. Compound **28** was obtained by reducing the carboxylic group of compound **27** using lithium aluminum hydride in THF. This was followed by the conversion of compound **28** to its corresponding azido derivative **29** through a reaction with sodium azide, as shown in **Figure 7**. The reaction of azide derivative **29** with *N*,*N′*-disuccinimidyl carbonate in an alkaline medium gave derivative **30**. Hydrogelator **32** was then synthesized by reacting compounds **30** with *L*-phenylalanyl-*L*-phenylalanine **31** in a solvent mixture of THF: H_2O in the presence of Hünig's base. Compound **32** forms a hydrogel in a mixture of H_2O: DMSO (5% DMSO) at a pH of 3.7. The CGC of hydrogelator **32** was reported to be 0.02–0.1 wt%.

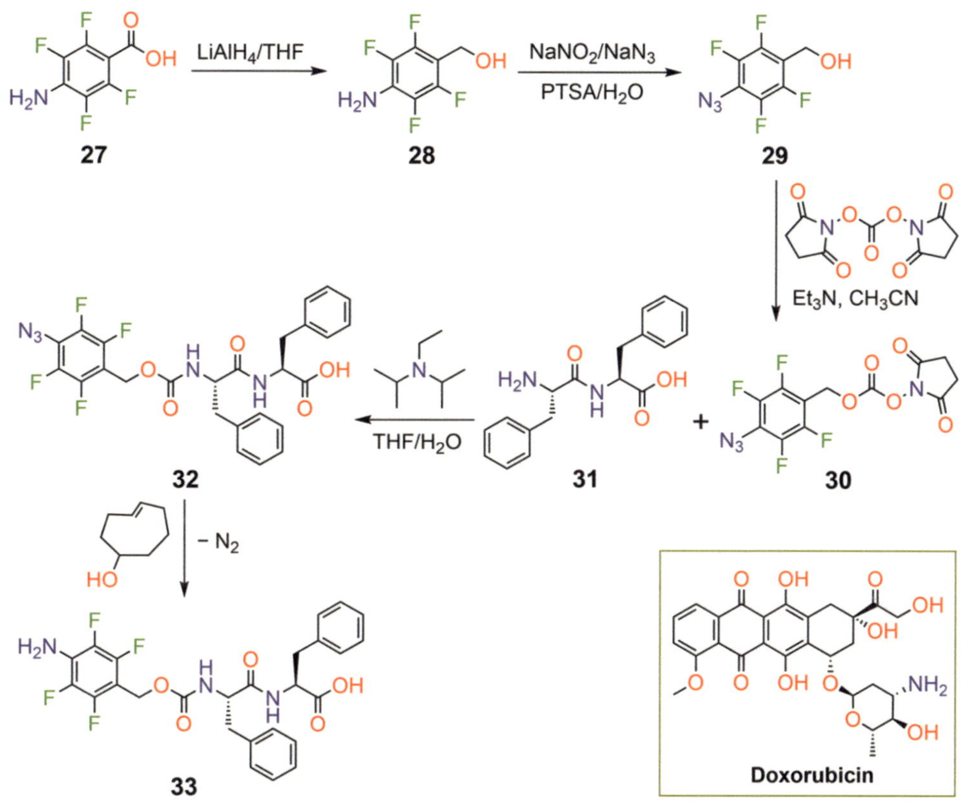

Figure 7.
Synthetic pathway of hydrogelator **32** *and the chemical structure of the drug molecule [38].*

Hydrogelator **32** assembles into entangled fibrous network structures with the aid of π-π stacking interactions. After 4 hours at 37°C, the hydrogel experienced a gel-to-sol transition by treating with 5 mM of the biorthogonal reagent cyclooct-4-enol to afford dissolution product **33** [38]. The evaluation of the drug delivery capability of hydrogelator **32** was undertaken by encapsulating it with DOX at 0.1 wt%. The drug-releasing potential was accomplished by treating the drug-loaded hydrogel with 1 mM of the biorthogonal reagent cyclooct-4-enol. The anticancer drug DOX was released from the gel matrix (89%) after 24 hours of incubation at 37°C.

The reaction of the nucleoside guanosine **34** with 1-naphthaleneboronic acid **35** in an aqueous KOH solution forms the G-quadruplex hydrogel **36** (**Figure 8**) [39]. Hydrogelator **36** self-assembles through augmented interactions such as hydrogen bonding, π-π stacking, cationic, and ion-dipole interactions. In addition to the guanosine moiety, hydrogelator **36** incorporates pH-responsive boronate ester rings. The hydrogel forms nanofibers at physiological pH (pH 7.4), making it an ideal drug delivery system. The hydrogelator was found to be non-toxic and biocompatible at high doses when tested against HeLa, MCF-7 breast cancer, and HEK293 embryonic kidney cell lines [39]. The hydrogel's drug delivery potential was assessed using vitamin B12, vitamin B2, and the anticancer drug DOX, which were separately encapsulated within the hydrogel matrix during the preparation process. The release of the vitamin or drugs was achieved at a physiological pH of 7.4 and a temperature of 37°C.

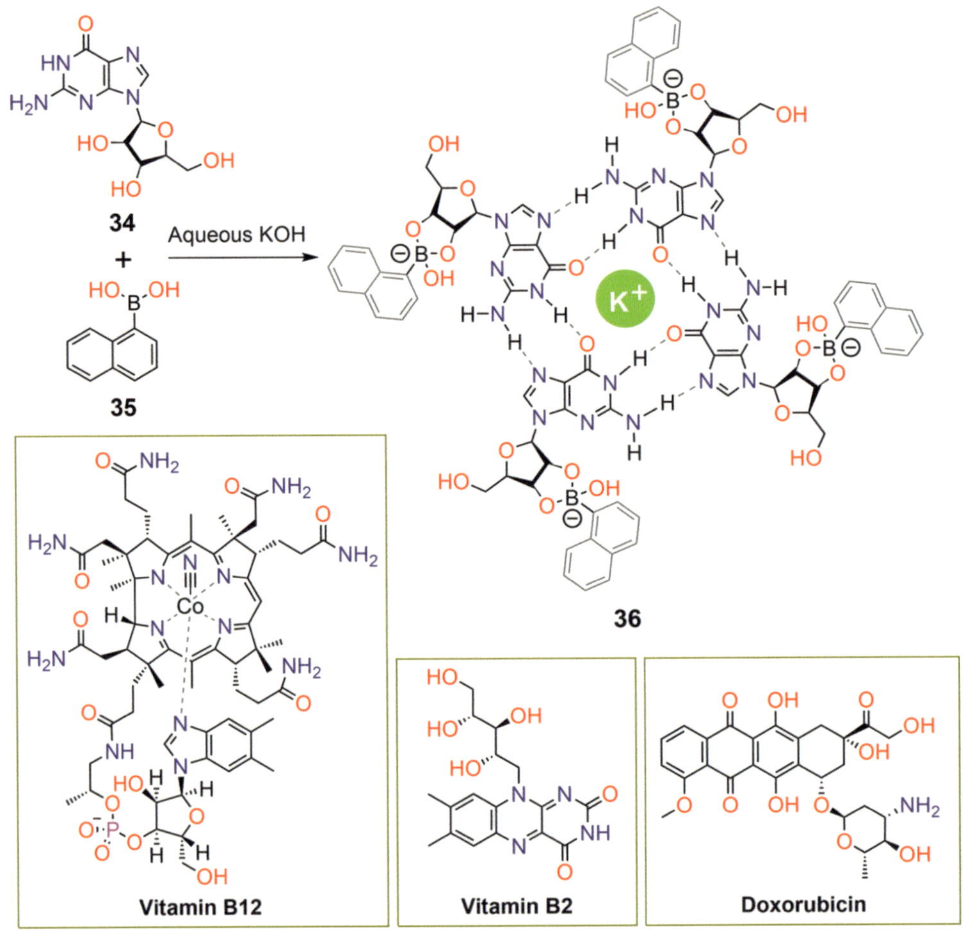

Figure 8.
Synthetic pathway of hydrogelator **36** *and the structures of the drug molecules [39].*

The hydrogel matrices released 30% and 60% of vitamin B12 and vitamin B2 after 40 hours, respectively, and the complete release of all entrapped vitamins was observed after 94 hours. At an acidic pH of 4.8, vitamin B12 and vitamin B2 were released in amounts of 82% and 75%, respectively. At pH 7.4 and 4.8, the DOX drug achieved 40% and 76% release, respectively. The pH-responsive behavior of the hydrogel was attributed to the breakage of hydrogen bonding interactions and boronate ester bonds.

Nayak *et al.* described the synthesis of amphiphilic compound **41** as presented in **Figure 9** [40]. Alkylation of phenol **37** with bromo-alcohol **38** in an alkaline medium yielded alkylated derivative **39**. Compound **39** was reacted with maleic anhydride **40** in dichloromethane (DCM) to form derivative **41**. Hydrogels were formed from compounds with (n = 8) and (n = 10) at CGCs of 1.6% and 1.3% w/v, respectively, by dissolving the compounds in phosphate buffer solutions at varied pH. Compound **41** incorporates phenoxy, carboxylic, ester, and alkyl groups that were engaged in gelation *via* π-π stacking, hydrogen bonding, and van der Waals interactions. Curcumin was incorporated into the hydrogel matrix by mixing it with the hydrogelator (n = 10) at the CGC in a phosphate buffer solution (pH = 8). Drug release occurred *via*

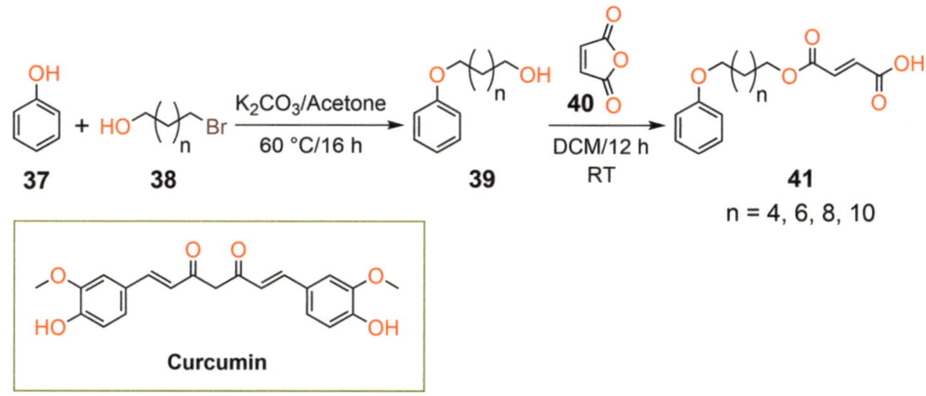

Figure 9.
Synthetic pathway of hydrogelator **41** *and the chemical structure of curcumin [40].*

enzymatic- and pH-triggered mechanisms. In the former, the drug-loaded hydrogel was treated with a solution of lipozyme, which cleaved the functional ester group, transforming the hydrogel into a sol state. In the latter, acidic treatment (pH = 3.5) of the drug-loaded hydrogel mediated persistent drug release.

Huang *et al*. reported the synthesis of the photoresponsive hydrogelator **57** *via* the synthetic steps demonstrated in **Figure 10** [41]. Amino-chromenone **42** reacted with bromoacetate **43** in acetonitrile under basic conditions to produce compound **44**. Derivative **44** was hydrolyzed under acidic conditions with TFA in a mixture of DCM and H_2O to yield carboxylic acid-functionalized derivative **45**. Coupling **45** with *D*-phenylalanine methyl ester hydrochloride **46** using 2-(1*H*-benzotriazole-1-yl)-1,1,3,3-tetramethylaminium tetrafluoroborate (TBTU) yielded the corresponding ester **47**, which was hydrolyzed to afford derivative **48**, with a free carboxylic group. The protection of the glycylglycine amino group using di-*tert*-butyl dicarbonate **50** yielded the (*tert*-butoxycarbonyl)glycylglycine **51**. The coupling reaction of derivative **51** with the protected boronic acid **54** afforded **55**, which on hydrolysis gave **56**. The coupling reaction of derivatives **48** and **56** with subsequent hydrolysis yielded gelator **57**. Coumarin-based derivative **57** formed a hydrogel in a 1: 2 mixture of polyethylene glycol 200 (PEG200) and water at a CGC of 2.7 mg/mL. The hydrogel assembled into entangled spiral-nanofibers with the aid of intermolecular hydrogen bonding and π-π stacking interactions. Hydrogelator **57** was found to exhibit a gel-to-sol transition upon exposure to UV irradiation at a wavelength of 365 nm, which was attributed to the breaking of the C-N bond. This property was exploited to utilize compound **57** as a drug carrier for the antineoplastic drug cytarabine hydrochloride. Thus, upon exposure to UV irradiation at 365 nm, the drug was released in a sustained and slow manner in a phosphate buffer with a pH range of 2.48–8.5. The rate of drug release was further enhanced by employing a laser as the irradiation source.

DuttKonar *et al*. reported a simple synthetic method of hydrogelators **60** and **63** [42]. Benzyl carbonochloridate **58** reacted with phenylalanine **59** in a basic medium to give gelator **60**. Similarly, the reaction of (9*H*-fluoren-9-yl)methyl carbonochloridate **61** with 3-aminobenzoic acid **62**, using the same conditions, resulted in the formation of hydrogelator **63**, as demonstrated in **Figure 11**.

Gelators **60** and **63** were utilized to construct fibrous network aggregates *via* hydrogen bonding and π-π stacking interactions. These network structures were formed at low CGCs of 0.02% and 0.05% *w/v*, respectively. Furthermore, when derivatives **60**

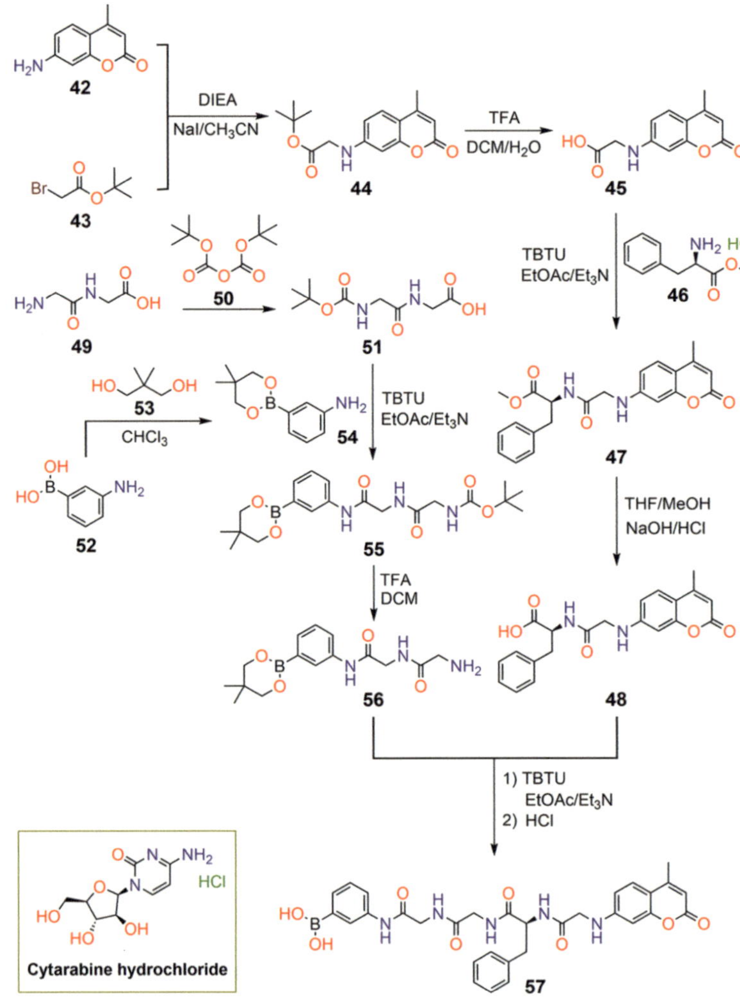

Figure 10.
Synthetic pathway of hydrogelator **57** *and the structure of the drug molecule [41].*

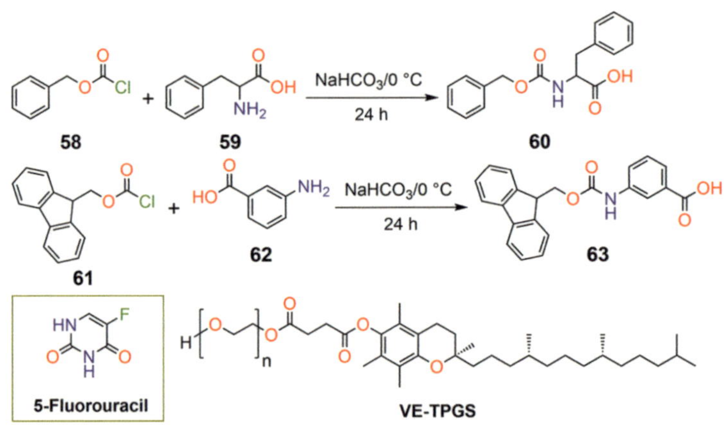

Figure 11.
Synthetic pathway of hydrogelators **60** *and* **63** *and the chemical structure of the drug molecule [42].*

and **63** were mixed with Vitamin E-TPGS in light paraffin oil, they formed nanoparticles. These nanoparticles were used as drug delivery systems for the anticancer drug 5-fluorouracil. The nanoparticles of compounds **60** and **63** exhibited high entrapment efficiencies of 77.57% and 84.52%, respectively. Drug release was assessed by immersing the drug-loaded nanoparticles in a phosphate buffer at a pH of 7.4. The drug release was pH-dependent, with hydrogelator **60** exhibiting a faster release rate than hydrogelator **63**. After 3 hours of immersion, hydrogelator **60** released 50% of the drug, whereas hydrogelator **63** released the same amount after 4 hours.

Bajaj *et al*. reported the synthesis of gelator **68** which incorporates naphthyl, amide, and hydrazide moieties as demonstrated in **Figure 12**. The synthetic method includes the reaction of 2-(naphthalen-1-yl)acetic acid **64** with methyl *L*-alaninate **65** in dry DCM in the presence of 1,1′-carbonyldiimidazole **66** to afford methyl (2-(naphthalen-1-yl)acetyl)-*D*-alaninate **67**. The reaction of ester **67** with hydrazine monohydrate in methanol yields hydrazide **68** [43]. The hydrogelation of compound **68** is mainly facilitated by intermolecular hydrogen bonding of the hydrophilic and π-π stacking of the hydrophobic constituents, potentially leading to the formation of nanofiber aggregates at a CGC of 1.0% *w*/*v* [43]. Hydrogelator **68** was employed as a drug delivery carrier for a series of anticancer medications (**Figure 12**). The drugs were added to saturated aqueous solutions of gelator **68** on hot, with maximum encapsulation efficiency achieved by 5-fluorouracil. The drug-loading potential of hydrogelator **68** is highly attributed to its ability to establish stable noncovalent interactions with the drug molecules.

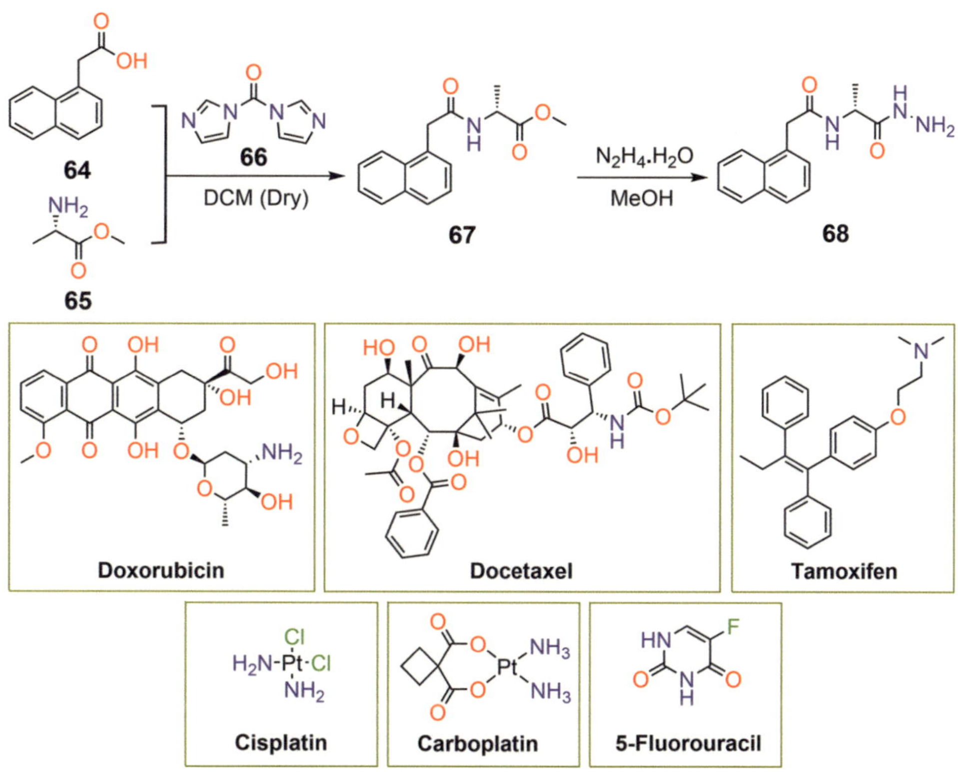

Figure 12.
Synthetic pathway of hydrogelator **68** *and the structures of the drug molecules [43, 44].*

Cisplatin and carboplatin showed minimum loading efficiency despite their relatively low molecular weights due to the absence of π-π stacking interactions and disruption of hydrogen bonding of the gelator by the amino groups [44]. The hydrazide group existing in the structural framework of gelator **68** could interact with DOX, resulting in high entrapment potential; however, this might lead to undesired structural modulation of the drug [44]. The release of the drugs was achieved at 37°C and a physiological pH. The drug-loaded hydrogels exhibited a release of 60–80%, except for cisplatin and carboplatin, which could be ascribed to the strong, established hydrogen bonding interactions with the hydrogelator.

Compound **71**, incorporating two fluorenyl moieties, was prepared by reacting (9*H*-fluoren-9-yl)methyl carbonochloridate **69** with 2,3-diaminopropanoic acid hydrochloride **70** in a basic medium as shown in **Figure 13** [45]. Shanmugam *et al.* demonstrated the ability of compound **71** to form stable hydrogel networks [46]. Compound **71** featured carbamate and fluorenyl groups, which facilitate the formation of intermolecular hydrogen bonding and π-π stacking interactions. The gelation of compound **71** occurred by dissolving it in DMSO with the successive addition of buffer solutions at varying pH values (pH = 4.9, 7.4, and 9.1) to form entangled nanofibers. The CGCs at pH values of 4.9, 7.4, and 9.1 were reported to 0.3, 1.0, and 1.3 wt%, respectively. The drug encapsulation and release potential of hydrogel **71** was assessed at pH 7.4 using vitamin B12. The hydrogel prepared at a gelation concentration of 1.5 wt% was diluted with the phosphate buffer containing vitamin B12. The gel released 50% of vitamin B12 in a sustained and slow fashion after 24 hours of incubation. Hydrogelator **71** revealed cell viability and proliferation against the human fibroblasts (3T3) cell line utilizing the MTT assay, indicating its biocompatibility. Nilsson *et al.* reported the preparation of compounds **75**–**77** containing fluorenyl moieties (**Figure 14**) [47]. Compound **72** reacted with the protected *tert*-butyl(3-aminopropyl)carbamate **73** in DCM: DMF in the presence of EDC, HOBt, and Hünig's base to afford **74**. Acidic hydrolysis of compound **74** with TFA in DCM yielded derivatives **75**–**77**. Gelation took place by treating heated

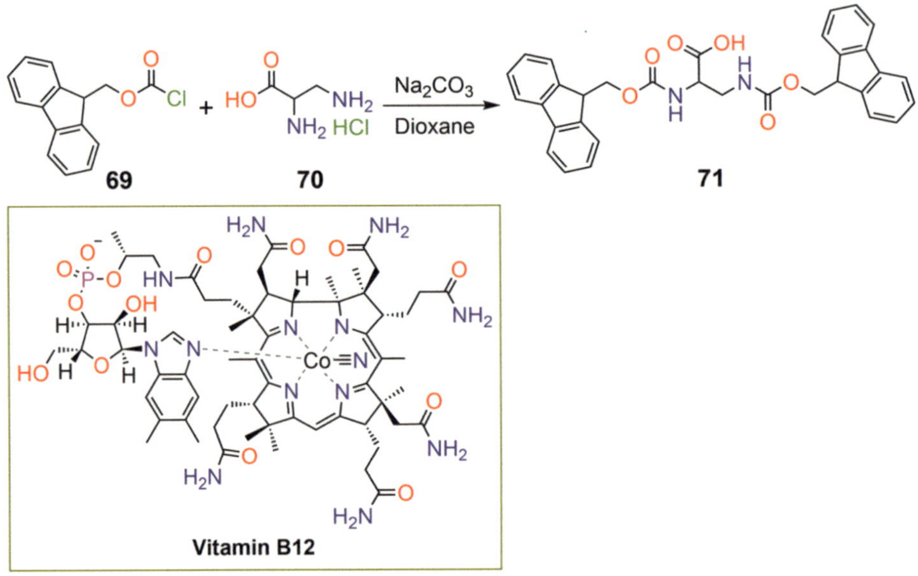

Figure 13.
Synthetic pathway of hydrogelator 71 and the chemical structure of the drug molecule [45, 46].

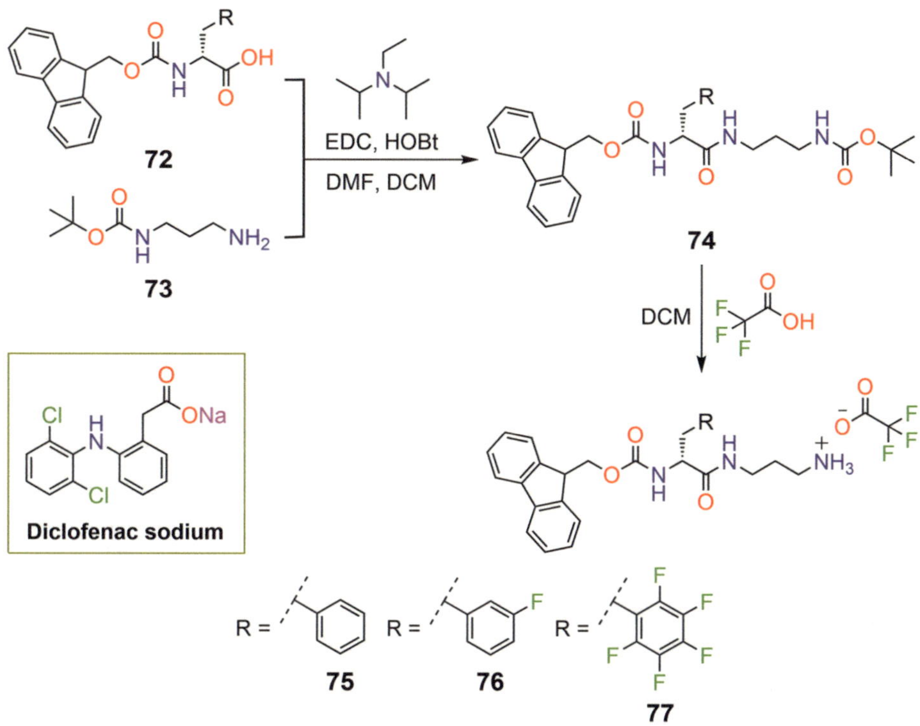

Figure 14.
Synthetic pathway of hydrogelators 75–77 and the structure of the drug molecule [47, 48].

aqueous solutions of **75–77** with NaCl at varied concentrations (2.5 mM, 5 mM, and 10 mM). The addition of NaCl boosts gelation by decreasing the repulsive effect of the cationic ($-NH_3^+$) group. Compounds **75–77** assemble into nanofibers at or below a concentration of 10 mM [48].

Gelators **75–77** developed at or above 30 mM exhibited outstanding stability, rendering them appealing candidates for drug delivery applications. Gelators **75–77**, as well as an equimolar co-assembly of compounds **75** and **77**, were studied *in vitro* and *in vivo* for their ability to serve as drug delivery vehicles for the anti-inflammatory drug diclofenac. Diclofenac was dissolved in water and mixed with **75–77** or **75:77**, with the subsequent addition of NaCl to promote hydrogelation.

The drug-loaded mixtures were incubated with a phosphate-buffered saline solution at a pH of 7 and a temperature of 37°C. All the hydrogels displayed slow and sustained drug release over the course of time in the order **75:77** > **75** > **76** > **77**. When the drug-loaded gel **77** was injected into mice with localized induced injuries, the drug release was slow and sustained [48].

Drug loading can also be achieved by linking drug molecules to the gelators, while release can be triggered using appropriate stimuli. Kundu *et al.* described the synthesis of hydrogelator **87**, tethered to the anticancer drug 5-fluorouracil (**Figure 15**) [49]. The (*tert*-butoxycarbonyl)phenylalanine **78** reacted with ethyl phenylalaninate **79** in the presence of the coupling agent TBTU and Hünig's base at room temperature to yield **80**, which on hydrolysis gave **81**.

The reaction of the anticancer drug 5-fluorouracil **82** with the protecting agent *N,O-bis*(trimethylsilyl)acetamide **83** affords derivative **84**, which reacted with

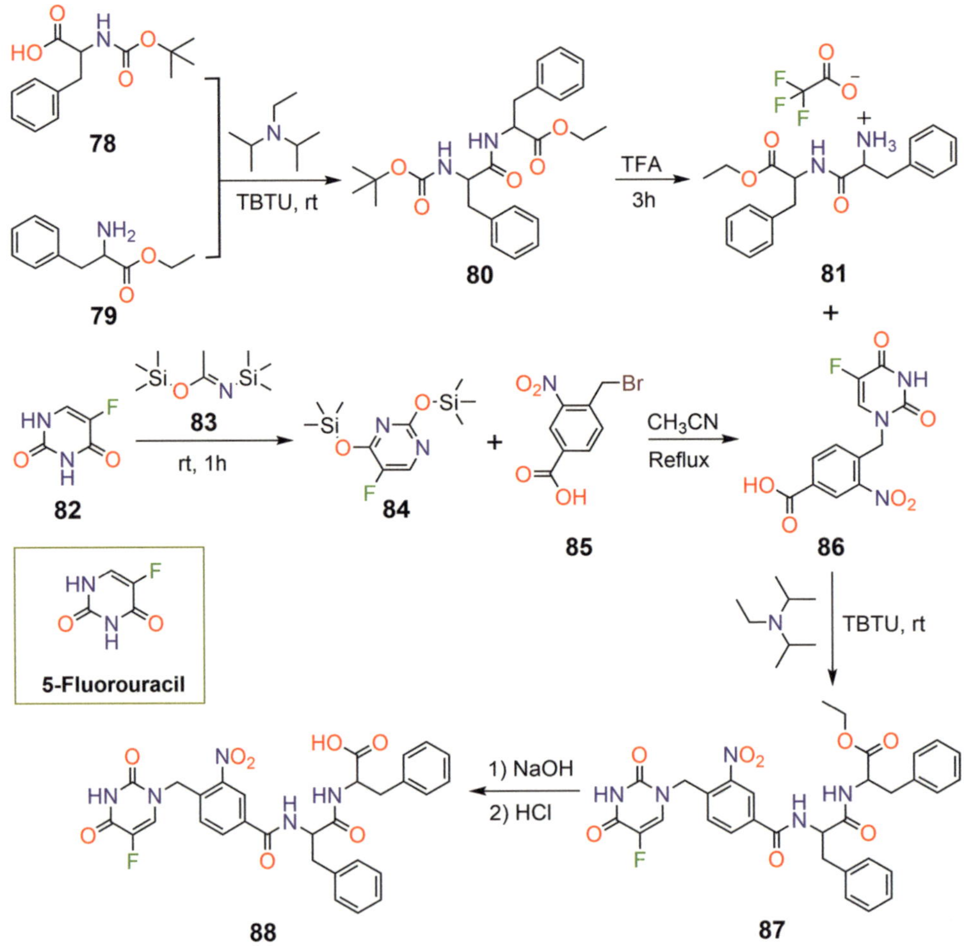

Figure 15.
Synthetic pathway of hydrogelator **88** *and the structure of the drug molecule [49].*

4-(bromomethyl)-3-nitrobenzoic acid **85** in acetonitrile under reflux to give **86**. The coupling of compound **81** with **86** forms the peptide-containing drug **87**, which on hydrolysis affords gelator **88**. The CGC of gelator **88** was determined to be 2 wt%. The surface morphology of hydrogelator **88** revealed the formation of entangled fibrils. The irradiation of gelator **88** using UV light at 365 nm induced 26% liberation the drug after 40 minutes [49]. The biocompatibility of hydrogelator **88** was investigated using the HeLa cell lines and the MTT assay, which revealed cell viability up to 111 µg.

Pianowski *et al*. reported the synthesis of hydrogelator **94**, which incorporates a photoswitchable azo functional group (–N=N–) as indicated in **Figure 16** [27]. The reaction of derivative **89** with *N*-Boc-*L*-lysine methyl ester hydrochloride **90** in DMF in the presence of TBTU and *N*-ethyl-*N*-isopropylpropan-2-amine gave **91**. Acidic hydrolysis of derivative **91** using TFA in DCM gave **92**. The reaction of **92** with *N*-methylmorpholine **93** in a solvent mixture of 2-butanol: AcOH and in the presence of Hünig's base afforded gelator **94** in the *trans*-form.

Compound **94** gelates in a phosphate-buffered saline solution (pH 7.4) in the form of fibril structures with a CGC of 3.0 wt%. The hydrogelation is triggered by hydrogen

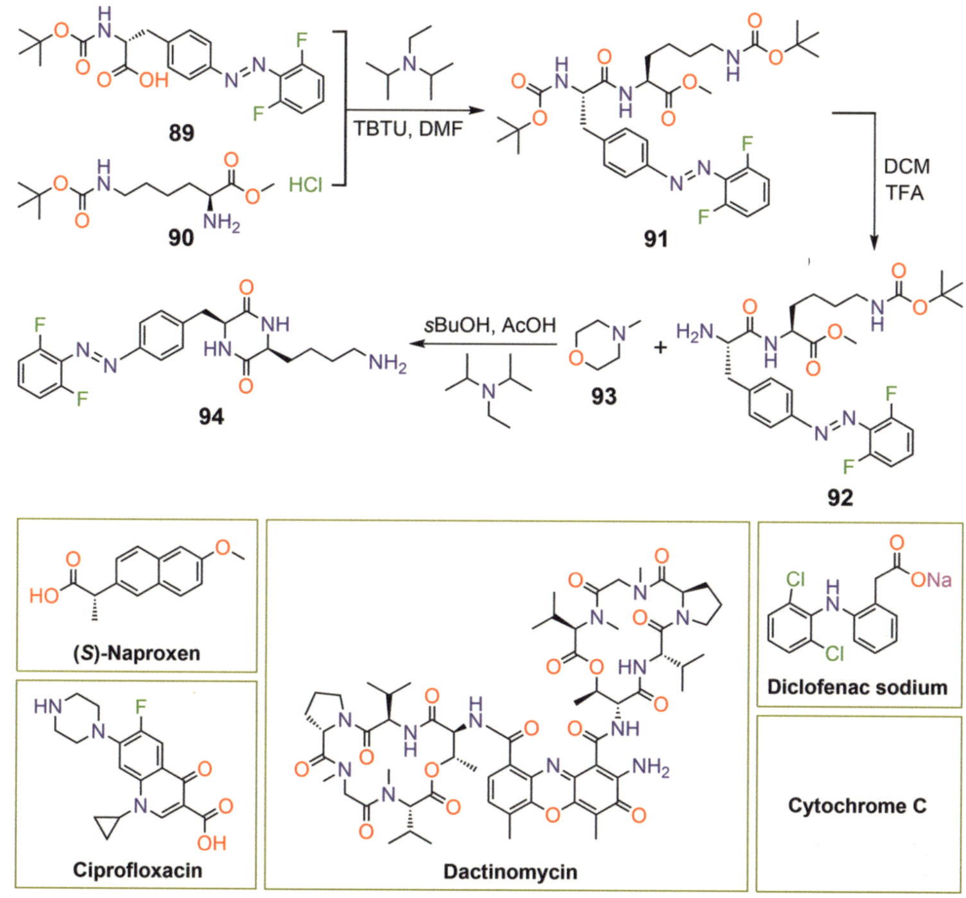

Figure 16.
Synthetic pathway of hydrogelator **94** *and the structures of the drug molecules [27].*

bonding of the amide functional motifs and π-π stacking interactions of the aryl moieties. The hydrogel underwent gel-to-sol transition upon exposure to green light at 523 nm after 3 hours of irradiation [27]. Several pharmaceutical molecules and a protein were encapsulated within the hydrogel matrix during fabrication, including (*S*)-naproxen, ciprofloxacin, dactinomycin, diclofenac, and cytochrome C. The cargo release was accomplished upon irradiation of the drug or protein-loaded hydrogel with a green light at 523 nm. The half maximal effective concentration (EC_{50}) of hydrogelator **94** against the HeLa cell lines using the MTT assay was >500 μM, indicating lower toxicity on the human cell lines [27]. In another related study, the photoresponsive hydrogelator **98** was designed as a drug delivery carrier for the anticancer drug doxorubicin (**Figure 17**) [50]. The reaction of compound **95** with *N*-Boc-*L*-lysine methyl ester hydrochloride **90** in DMF in the presence of TBTU and Hünig's base featured the azo-derivative **96**.

Hydrolysis of compound **96** with TFA in DCM afforded compound **97**, which, on reaction with *N*-methylmorpholine **93** in 2-butanol: AcOH and Hünig's base yielded the corresponding derivative **98**. Similar to fluorinated hydrogel **94**, hydrogelator **98** is assembled into fibers by hydrogen bonding and π-π stacking interactions. The gel-to-sol transition was triggered by irradiation with UV light at 365 nm, while the sol-to-gel transition was achieved by irradiation with blue light at 460 nm. The

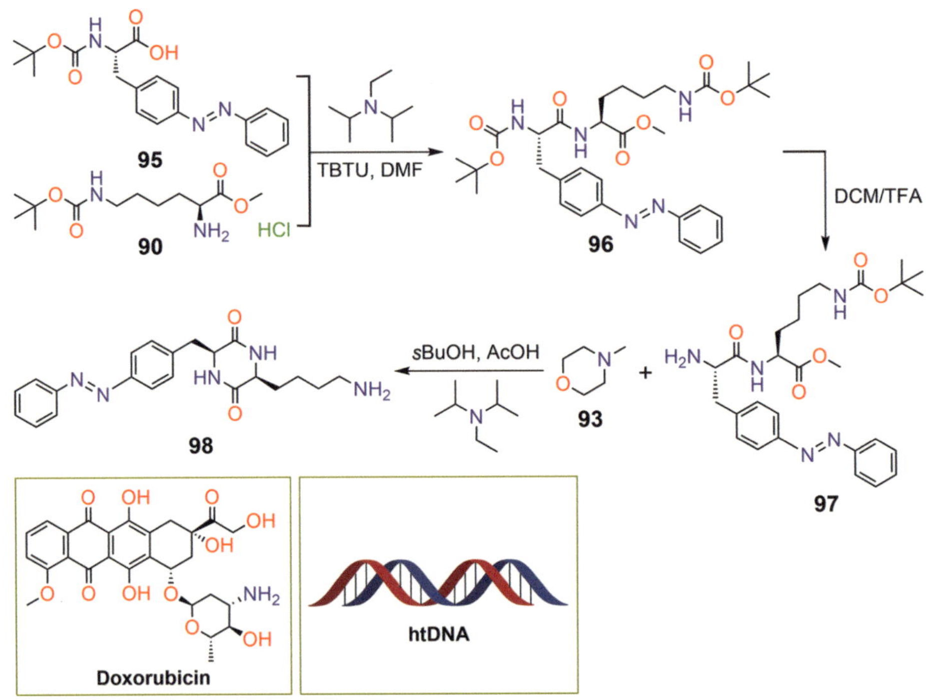

Figure 17.
Synthetic pathway of hydrogelator **97** *and the structure of the drug molecule [50].*

irradiation of hydrogelator **98** with UV light at 365 nm disrupted intermolecular noncovalent interactions with transformation to the liquid *cis*-form. The CGC of gelator **98** was determined to be 1.0% [50]. The mechanical properties of hydrogelator **98** were improved by conducting gelation in aqueous NaCl solution. Hydrogelator **98** was utilized as a delivery system of DOX and herring testis DNA (htDNA). The cargos were encapsulated with the hydrogel during the synthesis at 1.5% gel concentration. After irradiating the DOX- or htDNA-loaded hydrogel with UV radiation at 365 nm for 30 minutes, 70% and 40% of DOX and htDNA were released, respectively.

In summary, **Table 1** presents an overview of the hydrogelators discussed in this chapter, along with their CGCs, the substances that can be trapped within the hydrogel matrix, and the conditions under which drug release occurs.

Entry	CGC	Cargos	Stimulus	References
3	—	DOX	pH	[28, 29]
7	0.05% *w/v*	5-FU	pH	[30]
13	1 mg mL^{-1}	(*S*)-NAP, CAF	pH	[32]
16	4.7% *w/v*	IM	Franz diffusion	[34]
21, 22	0.01% *w/v*	5-FU, DOX	pH	[35]
26	—	CUR	pH	[37]
32	0.02–0.1 wt%	DOX	Bioorthogonal trigger	[38]
36	—	B12, B2, DOX	pH	[39]

Entry	CGC	Cargos	Stimulus	References
41	1.3 *w/v*	CUR	Enzyme/pH	[40]
57	2.7 mg mL^{-1}	ARA-C	Light	[41]
60	0.02% *w/v*	5-FU	pH	[42]
63	0.05% *w/v*	5-FU	pH	[42]
68	1.0% *w/v*	DOX, DTX, TAM, CP, CB, 5-FU	pH	[43]
71	0.3 wt%	B12	pH	[46]
75, 76, 77	2.5, 2.5, 5 mM	DCF	pH	[47, 48]
88	2 wt%	5-FU	Light	[49]
94	3.0 wt%	(*S*)-NAP, CIP, ACTD, Cytc, DCF	Light	[27]
98	1.0%	DOX, htDNA	Light	[50]

Abbreviations: DOX, Doxorubicin; 5-FU, 5-Fluorouracil; (S)-NAP, (S)-Naproxen; CAF, Caffeine; IM, Imatinib mesylate; CUR, Curcumin; B12, Vitamin B12; B2, Vitamin B2; ARA-C, Cytarabine; DTX, Docetaxel; TAM, Tamoxifen, CP, Cisplatin; CB, Carboplatin; DCF, Diclofenac; CIP, Ciprofloxacin; ACTD, Dactinomycin; Cytc, Cytochrome c.

Table 1.
Overview of hydrogelators, CGCs, drug trapping, and release conditions.

5. Conclusions and prospects

LMWHs have emerged as promising materials with potential applications in various research areas. The unique properties of LMWHs, such as cost-effective preparation, easy functionalization, biocompatibility, low toxicity, and ability to respond to external stimuli, make them ideal candidates for different biomedical applications. LMWHs have shown great potential as drug delivery vehicles for a variety of drugs. However, there are several obstacles that should be addressed in order to fully exploit their promise in drug delivery applications. One of the major drawbacks of LMWHs is their weak mechanical strength, which can be tuned by modifying the structure of hydrogelators, such as incorporating more functional motifs, to promote and strengthen noncovalent interactions. Co-assembly can also reinforce the mechanical properties of LMWHs. The stability of LMWHs under physiological conditions or in response to external stimuli is a crucial factor that should be considered during the development of hydrogels intended for drug delivery applications. LMWHs possess limited drug-loading capacity and, in some circumstances, lack targeted drug delivery. Therefore, the development of more sophisticated strategies for improving the drug-loading potential of LMWHs is an urgent necessity. Moreover, drug release from LMWHs may also be inefficient, leading to suboptimal therapeutic effects. Thus, careful design of the structure of the hydrogelators and selection of the triggers are crucial for achieving optimal drug release. Finally, sustained and slow releases of drugs are two vital aspects to be accomplished for efficient drug delivery using LMWHs.

We hope that this chapter will spark further investigations into the synthetic approaches of LMWHs and lead to the discovery of novel gelators with potential applications in drug delivery.

Conflict of interest

The authors declare no conflict of interest.

Author details

Hany F. Nour[1*], Ahmed M. Salama[2], Badria H.A. Al-Dhuwayin[3] and Amal F. Seliem[3]

1 Photochemistry Department, Chemical Industries Research Institute, National Research Centre, Cairo, Egypt

2 State Key Laboratory of Chemical Resource Engineering, Beijing University of Chemical Technology, Beijing, China

3 Department of Chemistry, College of Science and Arts, Najran University, Najran, Saudi Arabia

*Address all correspondence to: hany.nour@daad-alumni.de

References

[1] Patil SP, Jeong HS, Kim BH. A low-molecular-weight supramolecular hydrogel of riboflavin bolaamphiphile for VEGF-siRNA delivery. Chemical Communications. 2012;**48**(71):8901-8903

[2] Patil SP, Kim S-H, Jadhav JR, Lee J-H, Jeon EM, Kim K-T, et al. Cancer-specific gene silencing through therapeutic siRNA delivery with B vitamin-based nanoassembled low-molecular-weight hydrogelators. Bioconjugate Chemistry. 2014;**25**:1517-1525

[3] Sedghiniya S, Soleimannejad J, Blake AJ. A low molecular weight Zr(IV) metallogel for protein delivery. Materials Today Communication. 2021;**27**:102448

[4] Jagrosse ML, Agredo P, Abraham BL, Toriki ES, Nilsson BL. Supramolecular phenylalanine-derived hydrogels for the sustained release of functional proteins. ACS Biomaterials Science & Engineering. 2023;**9**(2):784-796

[5] McCloskey AP, Lee M, Megaw J, McEvoy J, Coulter SM, Pentlavalli S, et al. Investigating the *in vivo* antimicrobial activity of a self-assembling peptide hydrogel using a galleria mellonella infection model. ACS Omega. 2019;**4**:2584-2589

[6] ReddySM,ShanmugamG,DuraipandyN, Kiran MS, Mandal AB. An additional fluorenylmethoxycarbonyl (Fmoc) moiety in di-Fmoc-functionalized L-lysine induces pH-controlled ambidextrous gelation with significant advantages. Soft Matter. 2015;**11**(41):8126-8140

[7] Bernet A, Behr M, Schmidt H-W. Supramolecular hydrogels based on antimycobacterial amphiphiles. Soft Matter. 2012;**8**:4873-4876

[8] Cross ER, Coulter SM, Fuentes-Caparros AM, McAulay K, Schweins R, Laverty G, et al. Tuning the antimicrobial activity of low molecular weight hydrogels using dopamine autoxidation. Chemical Communications. 2020;**56**(58):8135-8138

[9] Zhao X, Zhang H, Gao Y, Lin Y, Hu J. A simple injectable moldable hydrogel assembled from natural glycyrrhizic acid with inherent antibacterial activity. ACS Applied Bio Materials. 2020;**3**(1):648-653

[10] Aldilla VR, Chen R, Martin AD, Marjo CE, Rich AM, Black DS, et al. Anthranilamide-based short peptides self-assembled hydrogels as antibacterial agents. Scientific Reports. 2020;**10**(1):770

[11] Zhao Q, Zhao Y, Lu Z, Tang Y. Amino acid-modified conjugated oligomer self-assembly hydrogel for efficient capture and specific killing of antibiotic-resistant bacteria. ACS Applied Materials & Interfaces. 2019;**11**(18):16320-16327

[12] Rizzo C, Arrigo R, Dintcheva NT, Gallo G, Giannici F, Noto R, et al. Supramolecular hydro- and ionogels: A study of their properties and antibacterial activity. Chemistry - A European Journal. 2017;**23**(64):16297-16311

[13] Tang Q, Plank TN, Zhu T, Yu H, Ge Z, Li Q, et al. Self-assembly of metallo-nucleoside hydrogels for injectable materials that promote wound closure. ACS Applied Materials & Interfaces. 2019;**11**(22):19743-19750

[14] Wei Q, Chang Y, Ma G, Zhang W, Wang Q, Hu Z. One-pot preparation of double network hydrogels via enzyme-mediated polymerization and

post-self-assembly for wound healing. Journal of Materials Chemistry B. 2019;**7**(40):6195-6201

[15] Liang C, Zhang L, Zhao W, Xu L, Chen Y, Long J, et al. Supramolecular nanofibers of drug-peptide amphiphile and affibody suppress HER2+ tumor growth. Advanced Healthcare Materials. 2018;**7**(22):e1800899

[16] Citossi F, Smith T, Lee JB, Segal J, Gershkovich P, Stocks MJ, et al. Self-assembling benzothiazole-based gelators: A mechanistic understanding of *in vitro* bioactivation and gelation. Molecular Pharmaceutics. 2018;**15**(4):1578-1586

[17] Bhagat SD, Srivastav A. Amphiphilic phenylalanine derivatives that temporally generate reactive oxygen species from water in the presence of Au(III) ions. Biomaterials Science. 2020;**8**(17):4750-4756

[18] Zanna N, Focaroli S, Merlettini A, Gentilucci L, Teti G, Falconi M, et al. Thixotropic peptide-based physical hydrogels applied to three-dimensional cell culture. ACS Omega. 2017;**2**(5):2374-2381

[19] Tang JD, Mura C, Lampe KJ. Stimuli-responsive, pentapeptide, nanofiber hydrogel for tissue engineering. Journal of the American Chemical Society. 2019;**141**(12):4886-4899

[20] Suga T, Osada S, Narita T, Oishi Y, Kodama H. Promotion of cell adhesion by low-molecular-weight hydrogel by Lys based amphiphile. Materials Science and Engineering: C. 2015;**47**:345-350

[21] Talloj SKM, Mohammed M, Lin H-C. Construction of self-assembled nanostructure-based tetraphenylethylene dipeptides: Supramolecular nanobelts as biomimetic hydrogels for cell adhesion and proliferation. Journal of Materials Chemistry B. 2020;**8**(33):7483-7493

[22] Dessane B, Smirani R, Bougueon G, Kauss T, Ribot E, Devillard R, et al. Nucleotide lipid-based hydrogel as a new biomaterial ink for biofabrication. Scientific Reports. 2020;**10**(1):2850

[23] Chu NT, Chakravarthy RD, Shih NC, Lin YH, Liu YC, Lin JH, et al. Fluorescent supramolecular hydrogels self-assembled from tetraphenylethene (TPE)/single amino acid conjugates. RSC Advances. 2018;**8**(37):20922-20927

[24] Nour HF, El Malah T, Khattab TA, Olson MA. Template-assisted hydrogelation of a dynamic covalent polyviologen-based supramolecular architecture via donor-acceptor interactions. Materials Today Chemistry. 2020;**17**:100289

[25] Colquhoun C, Draper ER, Eden EG, Cattoz BN, Morris KL, Chen L, et al. The effect of self-sorting and co-assembly on the mechanical properties of low molecular weight hydrogels. Nanoscale. 2014;**6**(22):13719-13725

[26] Basak S, Singh I, Banerjee A, Kraatz H-B. Amino acid-based amphiphilic hydrogels: Metal ion induced tuning of mechanical and thermal stability. RSC Advances. 2017;**7**:14461-14465

[27] Karcher J, Pianowski ZL. Photocontrol of drug release from supramolecular hydrogels with green light. Chemistry - A European Journal. 2018;**24**(45):11605-11610

[28] Noteborn WEM, Vittala SK, Torredemer MB, Maity C, Versluis F, Eelkema R, et al. Switching the mode of drug release from a reaction-coupled low-molecular-weight gelator system by altering its reaction pathway. Biomacromolecules. 2023;**24**(1):377-386

[29] Poolman JM, Boekhoven J, Besselink A, Olive AG, van Esch JH, Eelkema R. Variable gelation time and stiffness of low-molecular-weight hydrogels through catalytic control over self-assembly. Nature Protocols. 2014;**9**(4):977-988

[30] Ghosh S, Nag S, Saha KD, Banerji B. *S*-benzyl cysteine based cyclic dipeptide super hydrogelator: Enhancing efficacy of an anticancer drug via sustainable release. Journal of Peptide Science. 2022;**28**(8):e3403

[31] Wadhavane PD, Gorla L, Ferrer A, Altava B, Burguete MI, Izquierdo MÁ, et al. Coordination behaviour of new open chain and macrocyclic peptidomimetic compounds with copper(ii). RSC Advances. 2015;**5**:72579-72589

[32] Valls A, Isabel Burguete MK, Kuret L, Altava B, Luis SV. Open chain pseudopeptides as hydrogelators with reversible and dynamic responsiveness to pH, temperature and sonication as vehicles for controlled drug delivery. Journal of Molecular Liquids. 2022;**348**:118051

[33] Bica K, Rijksen C, Nieuwenhuyzen M, Rogers RD. In search of pure liquid salt forms of aspirin: Ionic liquid approaches with acetylsalicylic acid and salicylic acid. Physical Chemistry Chemical Physics. 2010;**12**(8):2011-2017

[34] Kuddushi M, Patel NK, Rajput S, El Seoud OA, Mata JP, Malek NI. Temperature-responsive low molecular weight ionic liquid based gelator: An approach to fabricate an anti-cancer drug-loaded hybrid ionogel. ChemSystemsChem. 2020;**2**(5):e1900053

[35] Mehra RR, Basu A, Christman RM, Harjit J, Mishra AK, Tiwari AK, et al. Mechanoresponsive, proteolytically stable and biocompatible supergelators from ultra short enantiomeric peptides with sustained drug release propensity. New Journal of Chemistry. 2020;**44**:6346-6354

[36] Rodik RV, Anthony A-S, Kalchenko VI, Mélya Y, Klymchenko AS. Cationic amphiphilic calixarenes to compact DNA into small nanoparticles for gene delivery. New Journal of Chemistry. 2015;**39**:1654-1664

[37] Granata G, Petralia S, Forte G, Conoci S, Consoli GML. Injectable supramolecular nanohydrogel from a micellar self-assembling calix[4]arene derivative and curcumin for a sustained drug release. Materials Science and Engineering: C. 2020;**111**:110842

[38] Dadhwal S, Fairhall JM, Hook S, Gamble AB. Tetrafluoroaryl azide as an N-terminal capping group for click-to-dissolve diphenylalanine hydrogels. RSC Advances. 2020;**10**(16):9234-9244

[39] Ghosh T, Biswas A, Gavel PK, Das AK. Engineered dynamic boronate ester-mediated self-healable biocompatible G-quadruplex hydrogels for sustained release of vitamins. Langmuir. 2020;**36**(6):1574-1584

[40] Kumar BA, Nayak RR. Supramolecular phenoxy-alkyl maleate-based hydrogels and their enzyme/pH-responsive curcumin release. New Journal of Chemistry. 2019;**43**:5559-5567

[41] Liu Q, Wang H, Li G, Liu M, Ding J, Huang X, et al. A photocleavable low molecular weight hydrogel for light-triggered drug delivery. Chinese Chemical Letters. 2019;**30**:485-488

[42] Tiwari P, Rajagopalan R, Moin M, Soni R, Trivedi P, DuttKonar A. Can self-assembled hydrogels composed of aromatic amino acid derivatives function

as drug delivery carriers? New Journal of Chemistry. 2017;**41**:308-315

[43] Singh M, Kundu S, Reddy M, Sreekanth V, Motiani RK, Sengupta S, et al. Injectable small molecule hydrogel as a potential nanocarrier for localized and sustained *in vivo* delivery of doxorubicin. Nanoscale. 2014;**6**:12849-12855

[44] Gupta S, Singh M, Reddy M. A, Yavvari PS, Srivastava a, Bajaj a, interactions governing the entrapment of anticancer drugs by low-molecular-weight hydrogelator for drug delivery applications. RSC Advances. 2016;**6**:19751-19757

[45] Nishikawa T, Urabe D, Isobe M. Syntheses of *N*-acylisoxazolidine derivatives, related to a partial structure found in zetekitoxin AB, a golden frog poison. Heterocycles. 2009;**79**:379-385

[46] Arokianathan JF, Ramya KA, Janeena A, Deshpande AP, Ayyadurai N, Leemarose A, et al. Non-proteinogenic amino acid based supramolecular hydrogel material for enhanced cell proliferation. Colloids and Surfaces B: Biointerfaces. 2020;**185**:110581

[47] Rajbhandary A, Raymond DM, Nilsson BL. Self-assembly, hydrogelation, and nanotube formation by cation-modified phenylalanine derivatives. Langmuir. 2017;**33**(23):5803-5813

[48] Raymond DM, Abraham BL, Fujita T, Watrous MJ, Toriki ES, Takano T, et al. Low molecular weight supramolecular hydrogels for sustained and localized in vivo drug delivery. ACS Applied Bio Materials. 2019;**2**(5):2116-2124

[49] Das S, Horo H, Goswami U, Kundu LM. Synthesis of a peptide conjugated 5-fluorouracil gelator prodrug for photo-controlled release of the antitumor agent. ChemistrySelect. 2019;**4**(22):6778-6783

[50] Pianowski ZL, Karcher J, Schneider K. Photoresponsive self-healing supramolecular hydrogels for light-induced release of DNA and doxorubicin. Chemical Communications. 2016;**52**(15):3143-3146

Chapter 2

Self-Assembled Nanogels Consisting of Cholesterol-Bearing Polysaccharides and Their Applications in Medicine

Haddad Antonia-Nancy, Michiaki Matsumoto and Yoshiro Tahara

Abstract

Cholesterol-bearing polysaccharides form self-assembled nanogels in water, which are versatile materials suitable for numerous applications in medicine. They are used in cancer vaccines, nasal vaccines, gene delivery, and regenerative medicine. Self-assembled nanogels encapsulate and provide controlled release of various drugs, including proteins (antigens for vaccines and growth hormone for regenerative medicine) or genes (siRNA and plasmid DNA). Moreover, self-assembled nanogel cross-linked macro-sized gels can act as scaffolds to support cell growth and tissue regeneration, making them valuable in tissue engineering and bone repair. Overall, self-assembled nanogels have a variety of medicinal uses and special properties that can improve patient care and progress the medical field.

Keywords: nanogels, self-assembly, cancer vaccine, nasal vaccine, gene delivery, regenerative medicine

1. Introduction

Self-assembled nanogels are created when molecules, including polymers, self-assemble into three-dimensional cross-linked networks, which can entrap therapeutic compounds or medications. The development of self-assembled polymers dates back to the 1980s, when scientists first started looking into the characteristics of self-assembled micelles. Micelles have a core-shell structure using amphiphilic block copolymer molecules with hydrophilic and hydrophobic polymer segments [1]. Following the report of polymer micelles as self-assembled polymer-based nanoparticles, self-assembled nanogels were first reported in 1993 [2], and researchers started looking into the potential of self-assembled nanogels for drug delivery in the 1990s. Compared with alternative drug delivery systems, these materials have a number of advantages, including high stability, biocompatibility, and the capacity to encapsulate both hydrophobic and hydrophilic drugs. Novel varieties of nanogels have been

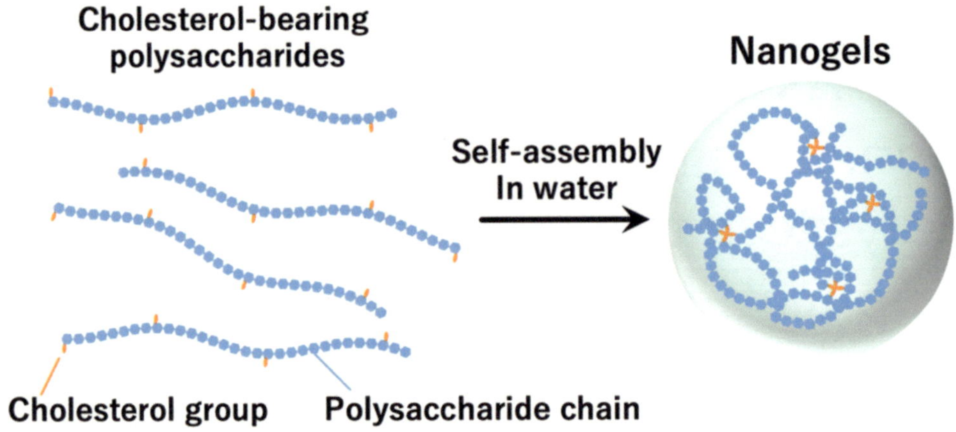

Figure 1.
Self-assembly of cholesterol-bearing polysaccharides.

created for various purposes, such as tissue engineering, medication delivery, and diagnostic imaging. In 2007, the International Union for Pure and Applied Chemistry (IUPAC) published a document defining the term "nanogel" [3], and in 2009, a detailed review of nanogels and medical applications was published [4]. Nanogels are regarded as one of the most important formulation classes of nanoparticles. Although various research articles and reviews on nanogels have been published, even in the 2020s, it should be noted that nanogels are different from other polymer-based nanoparticulate formulations, such as polymeric micelles, polymeric nanospheres, and polymersomes. The water content of nanogels (especially hydrogel nanoparticles) should be high because they are nanometer-sized hydro-"GELs." Cholesterol-bearing polysaccharides contain a high amount of water and are the most studied self-assembled nanogels. This review is focused on cholesterol-bearing polysaccharides, including their history and medical applications (**Figure 1**).

2. Basic research of self-assembled nanogels

2.1 Pioneer research of self-assembled nanogels

The research related to "supramolecular chemistry," "self-assembly," and "host-guest interactions" had been attracted much attentions in the research field of chemistry from 1980s to 2000s [5]. The biomedical applications had been carried out using these self-assembled materials including liposomes [6] and polymeric micelles [1]. In 1993, cholesterol-bearing pullulan (CHP) emerged as one of the first generation of self-aggregating nanogels [2]. Nanogels are nanometer-sized hydrogels. The self-aggregation and complexation behavior of nonionic pullulan derivative CHP-55-1.6, in which cholesterol groups were substituted at 1.6 per 100 glucose units of pullulan (molecular weight of 55,000), in water has been examined. CHP-55-1.6 generates monodispersive nanoparticles, and a single nanoparticle has microdomains comprising several hydrophobic cholesterol cores. The review published in 2009 described that this CHP nanogel is the firstly reported nanogel clearly showing the properties of nanometer-sized hydrogels [4]. In 1997, CHP nanogels with pullulan of various molecular weight pullulan and cholesteryl groups of various degrees of substitution were able to form monodisperse self-assembled nanoparticles in water. As the degree of substitution of the cholesteryl moiety increases, the size of the self-aggregate increases. Cholesteryl moieties are distributed within the particle to create a poly-core structure and to provide non-covalent cross-linking sites for the gels. Therefore, the hydrogel nanoparticle's overall size and polymer density are readily controlled by substitution degree of cholesteryl group [7].

2.2 Functions of molecular chaperone

Complexation between the CHP nanogel and bovine serum albumin (BSA) or insulin was investigated in 1996 and 1998 [8, 9]. These studies found that the complex of insulin and CHP nanogel is a stable colloid that can resist thermal denaturation and enzymatic degradation, thus maintaining its physiological activity. Subsequently, self-assembled nanogels have been investigated as molecular chaperones. Generally, proteins functioning as molecular chaperones are popular because they can avoid aggregation and misfolding. Owing to their high surface area and capacity to encapsulate proteins, self-assembled nanogels can imitate natural chaperones and function as synthetic molecular chaperones to prevent protein aggregation and misfolding in vitro. By creating a safe environment, nanogels can prevent denaturation of proteins. For example, proteins enclosed in the hydrophobic core of nanogels are protected from hostile environments.

In 1999, CHP nanogels were shown to undergo complexation with carbonic anhydrase B, which is then fully prevented from irreversibly aggregating upon heating. Upon release, the enzyme refolds to take on its natural form, recovering almost all of its original activity. The enzyme's heat stability is greatly enhanced by the unfolded form that is captured, allowing for efficient refolding [10]. Extensive studies on the use of CHP nanogels in protein refolding were reported in 2003. CHP nanogels act as molecular chaperones and efficiently prevent protein aggregation during the refolding of denatured proteins, such as carbonic anhydrase and citrate synthase. The proteins are released upon dissociation of the gel structure by cyclodextrins (CD), allowing for high enzyme activity. The nanogels trap refolded intermediate proteins and can be used in the renaturation of recombinant proteins from the serine protease family [11]. The dynamic properties of CHP-CD complexes, made of CHP and CD, have been investigated using various types of CDs. CHP nanogels self-regulate protein attachment and dissociation depending on the concentration of CD [12]. Another study examined the effect of CHP nanogels on the colloidal and thermal stability of lipase. The results showed that CHP nanogels significantly increase the enzyme activity and thwart the denaturation and aggregation of the lipase upon heating [13]. In another study, CHP nanogels were used to capture partially or completely unfolded green fluorescent protein. After complexation with methyl-beta-cyclodextrin, CHP nanogels separate to produce dissociated CHP, enabling the release and folding of green fluorescent protein. The folding kinetics are similar to that of spontaneous folding [14]. In a 2011 study, artificial chaperones based on CHP nanogels were used to enhance the folding efficiency of rhodanese and various water-soluble proteins generated in cell-free systems. In the presence of cyclodextrin, proteins fold correctly to produce enzymatically active proteins because the nanogel inhibits protein aggregation [15].

Medical applications have been explored. For example, in 2006, CHP nanogels were used to bind to partially unfolded proteins to prevent protein aggregation. CHP nanogels have been used as artificial chaperones to prevent the development of amyloid beta-protein (Ab) fibrils, which are thought to represent a crucial stage of Alzheimer's disease. These fibrils contain up to 6–8 Ab molecules, and amine-modified CHP (described below) has been shown to inhibit Ab aggregation [16].

2.3 Varieties of self-assembled nanogels

There are multiple types of self-assembled nanogels. The broad class of self-assembled nanogels can be created by using a wide range of building elements [17]. Research conducted in 2009 discovered that cholesterol modified with a highly branched cyclic dextrin (CH-CDex) can spontaneously form stable, monodisperse nanogels in water. These nanogels exhibit exceptional colloidal stability by capturing insulin. CH-CDex nanogels prevent insulin from dissociating for over 30 days, and in the presence of BSA, insulin is gradually released. A nanogel–protein complex was found to be more stable than the lone protein owing to its highly branched CDex structure [18]. In another study, a monodisperse, spherical, hyperbranched nanoparticle called enzyme-generated glycogen (ESG) was proven to serve as a synthetic chaperone for protein engineering. ESG has a cholesterol group, which enables the creation of amphiphilic ESG nanoballs that can form complexes with proteins. The cluster nanogels are then broken up by cyclodextrin [19]. In a subsequent study using ESG containing hydrophobic groups, amphiphilic ESG nanogels with a radius of 15 nm were found to inhibit carbonic anhydrase from irreversibly aggregating. Without causing the nanogels to separate, cyclodextrin triggers the release of carbonic anhydrase in its active state [20]. Anionic nanogels based on hyaluronic acid (HA) and containing cholesterol groups have been developed for protein delivery. These nanogels can bind to different kinds of proteins without denaturing them, and they can associate with salt to form an injectable hydrogel. According to one study, the level of recombinant human growth hormone (rhGH) in plasma can be preserved for a whole week through the in situ gel formulation of rhGH with HA nanogel, which is a useful way to create sustained protein release systems [21]. In a recent study, HA nanogels were used as molecular chaperones of antibodies to prevent heat denaturation, revealing that HA nanogels can encapsulate a high amount of antibodies and increase the activities of antibodies [22]. In 2017, cholesterol-bearing xyloglucan (CHXG) nanogel was created by incorporating multibranched polysaccharide with xylose and galactose side chains into CHXG nanogel. This nanogel is selectively internalized by hepatocytes via their cell-surface galactose receptors [23].

Recently, several biomedical applications have been reported using hybrid materials of CHP nanogels and inorganic materials, which showed the magnetic responses and named as magnetic nanogel chaperone. Caspase-3 (protease inducing apoptosis) was directly delivered to the target HeLa cells by the magnetic nanogel chaperone after they were guided there by a magnetic field [24]. Subsequent study indicated that saporin (anticancer proteins) can be magnetically steered using magnetic nanogel chaperone in vivo and lowered the tumor's size in an oral cancer model [25].

3. Cancer immunotherapy

Owing to their effectiveness in delivering antigens to immune cells and in triggering immunological responses, CHP nanogels have attracted interest in the fields of cancer vaccines and cancer immunotherapy [26]. In early studies, CHP nanogels were used to encapsulate stable, long-lasting antigens and thus create cancer vaccines. These vaccines can be made to induce certain cytotoxic T lymphocyte (CTL) responses against cancer cells. Recently, CHP nanogels have been used to deliver antigens to tumors directly and to stimulate immunological responses, with checkpoint inhibitors, against cancer cells. Self-assembled nanogel-based immunotherapeutic drug delivery against cancers is reviewed in this section.

In 1998, a novel hydrophobized polysaccharide nanoparticle formula, (cholesterol-bearing pullulan (CHP) and mannan (CHM) encapsulating human epidermal growth factor receptor 2 (CHP-HER2 and CHM-HER2)) was created to transport an oncogene HER2-containing epitope peptide to the major histocompatibility complex (MHC) class I pathway, generating CTLs and boosting humoral immunity against cancer. In mice, the CHP-HER2 and CHM-HER2 complexes successfully triggered total tumor rejection, in which cluster of differentiation 8 (CD8)-positive T cells were crucial in the effector phase of in vivo tumor rejection. These findings suggest that self-assembled nanogel vaccination is useful for both cancer treatment and prevention [27]. Research further showed that peptides HER2p63-71 and HER2p780-788 (an epitope peptide to the MHC class I pathway) encapsulated in CHP nanogels activated CD8-positive CTLs to act against HER2-positive malignancies, resulting in the total rejection of tumors in mice. This supports the possibility of using researched vaccines for treating and preventing cancer [28]. Subsequently, the use of human dendritic cells in immunotherapy was researched with complexes of CHP nanogels and target proteins (a shortened HER2 protein). These complexes were taken up by human dendritic cells, which then processed the complexes to present the HER2p63 peptide, activating and enhancing CD8-positive T cells with the used peptide's specificity [29]. In 2006, the results of a human clinical trial using CHP-HER2 vaccination were reported. All processes were carried out under good manufacturing practice (GMP) grade condition. Five patients who received the immunization showed HER2-specific CD8 and CD4-positive T cell immunological responses, indicating that the vaccine was both safe and efficient in eliciting an immune response against HER2-expressing tumors [30]. In a 2007 study, nine patients who received the CHP-NY-ESO-1 complex vaccine showed a rise in the responses of CD8 and CD4-positive T cells, as well as two dominant NY-ESO-1 regions. Local redness at the injection site was shown in all patients, while these reactions disappeared within three days. The safety of subcutaneous administration of CHP nanogel was confirmed [31]. In a research in which patients with malignancies expressing HER were subcutaneously administered vaccines, the CHP-HER2 vaccine induced CD8- and CD4-positive T cell immune responses specific for 146HER2. Moreover, fourteen individuals with baseline negative HER2-specific IgG antibodies had raised 146HER2 antigens. In patients who received the lone CHP-HER2 vaccine, the antibodies were not reactive until the third to sixth immunization. In contrast, in patients who received CHP-HER2 with the granulocyte-macrophage colony-stimulating factor vaccine, antibodies were detectable after the second or third immunization and peaked after the third or fourth round [32]. A 2014 basic study on mice suggested that targeting medullary macrophages is an efficient strategy to induce immunological responses with CHP nanogel-based vaccinations. CHP nanogels were used to encapsulate a synthetic long peptide antigen, which efficiently reached the draining lymph node. Furthermore, long peptides were detected in medullary macrophages but were not detected in other macrophages or dendritic cells. These vaccines significantly inhibited in vivo tumor growth in both prophylactic and therapeutic settings [33]. Researchers also investigated whether a cholesterol-bearing cluster dextrin and CHP nanogels could activate

CTLs and thus the T helper 1 (Th1) pathway. In comparison to lone antigen, both self-assembled nanogel vaccines activated CTLs to a greater extent and stimulated the generation of IgG2a antibodies. The distribution of antigens in the draining lymph nodes delivered by nanogels was evaluated, revealing productive interactions with particular groups of cross-presenting dendritic cells [34]. To improve interactions with antigen-presenting cells in the lymphatic system, CHP nanogels were modified to exhibit a net anionic charge by carboxyl group substitution. These nanogels showed effective antigen presentation and significantly increased adaptive immunity [35]. The possibility of combining the CHP nanogel vaccine with an immune checkpoint inhibitor was investigated. An antigen-loaded CHP nanogel with an anti-PD-1 antibody increased the effectiveness of cancer treatments [36]. The above vaccines based on self-assembled nanogels were all administered by subcutaneous injection. In 2019, an antigen with CHP nanogels were systemically administered by intravenous injection to investigate immune resistance in solid tumor models. The results revealed that CD11b and F4/80-expressing tumor-associated macrophages (TAMs) were particularly important for antigen presentation and immune resistance. By intravenously injecting a delivery system targeting TAMs, namely, antigen-loaded CHP nanogels and a Toll-like receptor agonist, antigen-presenting activity in the cancer microenvironment was boosted, and silence tumors were transformed to tumors sensitive to adaptive tumor-specific CTLs [37].

4. Cationic nanogels and its applications (nasal vaccine and gene delivery)

As shown above, early development of self-assembled nanogels for medical applications mainly used CHP nanogels, which have electrically neutral surface charges. Since 2005, cationic group-modified self-assembled nanogels have been prepared for various applications, including intracellular drug delivery, nasal administration, and gene delivery.

4.1 Cationic nanogels

In 2005, monodisperse hybrid nanoparticles were created using quantum dots (QDs) and CHP nanogels modified with amino groups ($CHPNH_2$) to provide novel carriers for intracellular labeling. Compared with traditional carriers, namely, cationic liposomes, $CHPNH_2$-QD complexes demonstrated an improved efficiency for the absorption by multiple human cells [38]. Subsequently, the delivery of intracellular proteins via a $CHPNH_2$ nanogel system was studied. Other traditional cationic drug carriers (cationic liposomes and protein transduction domain-based carriers) were less effective at internalizing HeLa cells in the presence of serum than $CHPNH_2$ nanogels [39]. In another study, $CHPNH_2$ nanogels were shown to provide effective delivery of proteins to myeloma cells and primary CD4-positive T lymphocytes. These cells are often resistant to protein transduction domain-mediated delivery because of inadequate heparan sulfate expression [40].

A different kind of cationic nanogel, namely, cholesterol-bearing cycloamylose modified with spermine (CH-CA-spe), was created for gene delivery. The spermine component enables the CH-CA-spe nanogel to carry genes. Moreover, the polyamine interacts with DNA or RNA molecules through electrostatic interactions to encourage condensation and prevent degradation.

4.2 Nasal vaccines

Owing to their special qualities, including positive charge, size, and stability, cationic nanogels have attracted considerable attention as prospective vaccine delivery systems for nasal administration. Compared with other administration routes, the nasal route has a number of benefits, including the capacity to trigger mucosal and systemic immune reactions, as well as being noninvasive and simple to use. In the nasal mucosa, cationic nanogels can effectively attach to and distribute vaccination antigens for uptake by dendritic cells and other antigen-presenting cells. The positive charge of cationic nanogels interacts with the negative charge of the mucosal surface, which improves the adherence and penetration of cationic nanogels into the nasal epithelium.

In a 2010 research, to evaluate the potential of cationic CHP (cCHP) nanogels for the nasal vaccine delivery system, a nontoxic fragment of *Clostridium botulinum* type-A neurotoxin (BoHc/A)-loaded cCHP nanogels was prepared. The system successfully elicited potent immunological responses, and it did not accumulate in the brain or olfactory bulbs [41]. Subsequently, pneumococcal surface protein A (PspA) was combined with cCHP nanogel to provide a safer and more potent vaccine against pneumococcal respiratory infections. This nasal vaccination made from PspA and cCHP nanogels offered defense against *Streptococcus pneumoniae* and inhibited bacterial invasion and colonization in respiratory tracts. Systemic and nasal mucosal Th17 responses were elicited, and PspA-specific antibodies were produced without harmful effects [42]. Next, cCHP nanogels containing PspA antigens were nasally administered to nonhuman primates to examine the safety and effectiveness of vaccines against pneumococcal infection. Compared with lone PspA, PspA-nanogel demonstrated longer-term retention in the nasal cavity. Productions of PspA-specific blood IgG and IgA antibodies were induced in cynomolgus macaques. Th2 and Th17 cytokine responses, as well as elevated levels of micro-RNA (miR)-181a in the blood and miR-326 in macaque respiratory tract tissues, mediated these immune responses, showing that PspA-nanogel nasal vaccination was safe and efficient for preventing pneumonia in human [43]. A trivalent vaccine using cCHP nanogels was developed to provide efficient immunity against all serotypes of *Streptococcus pneumoniae*. The vaccine was administered to macaques with a nasal spray device suitable for human use, and the immunogenicity and protective effectiveness of the vaccine were evaluated. PspA-specific antibodies were produced after nasal administration of the cCHP-trivalent PspA vaccine. The lungs of macaques had reduced lung inflammation and bacterial counts, indicating protection from pneumococcal infection [44]. The research focused on the absorption of the PspA-cCHP nanogel formulations by the nasal mucosa and the gradual release of PspA from mice epithelial cells. The immunologic activity and PspA content of the PspA-cCHP nanogel formulation were decreased by heat exposure. Mice injected with the trivalent PspA-cCHP nanogel formulation developed comparable amounts of IgG and IgA antibodies in their mucous membranes [45]. According to these studies, the PspA-cCHP nanogel is an effective and promising system for nasal vaccination. Ghrelin is a hormone that increases food intake and decreases energy expenditure. A nasal vaccine for obesity was developed to prevent the pain associated with injections and the negative and risky skin consequences of using ghrelin-PspA fusion protein. In mice with diet-induced obesity, intranasal administration of the ghrelin-PspA vaccine resulted in the production of serum IgG antibodies against ghrelin and decreased body weight. This impact is in part related to the increased expression level of mitochondrial uncoupling protein 1 in brown adipose tissue [46]. Recently, a nasal vaccine has been created to induce the generation of nontypeable *Haemophilus influenzae* (NTHi)-specific secretory IgA and thus inhibit the development of biofilms in the respiratory tract. NTHi surface antigen P6, which is conserved among 90% of NTHi strains, was added to a cCHP nanogel to create a nasal vaccine. Mice were nasally inoculated with the P6-cCHP nasal vaccine, which significantly induced the synthesis of P6-specific IgA antibodies in mucosal fluids, including nasal and middle ear washes [47].

4.3 Gene delivery

A specific kind of cationic nanogel, namely, cholesterol-bearing cycloamylose modified with spermine (CH-CA-spe), was created for gene delivery. The spermine component enables the CH-CA-spe nanogel to carry genes. Moreover, the polyamine interacts with DNA or RNA molecules through electrostatic interactions to encourage condensation and prevent degradation. In the first study, the abilities of CH-CA-spe nanogel and CA-spe to deliver siRNA were investigated. The CH-CA-spe nanogel showed higher gene silencing effect than CA-spe [48]. In 2014, a CH-CA-spe nanogel was tested as a vehicle for siRNA targeting vascular endothelial growth factor (VEGF) in tumor cells. Renal cell carcinoma (RCC) cells ingested the complex of siRNA and CH-CA-spe nanogel, which effectively knocked down VEGF. Intra-tumor injections of the compound drastically reduced RCC development and neovascularization in mice [49]. In another study, cycloamylose was modified with cholesterol and diethylaminoethane to create the CH-CA-DEAE nanogel, which was used to deliver unmethylated CpG oligodeoxynucleotides (immunostimulators). The CH-CA-DEAE nanogel successfully transported native CpG DNA to macrophage-like cells, causing the release of cytokines. Negative control oligonucleotides (with single mutation sample) with the CH-CA-DEAE nanogel prevented cytokine production, and phosphorothioate-modified CpG with the CH-CA-DEAE nanogel reduced cytokine production. These findings suggest that the CH-CA-DEAE nanogel is a potential nucleic acid adjuvant delivery system for native CpG DNA [50].

5. Nanogel tectonic materials and regenerative medicine

Nanogels can be cross-linked to create novel hydrogels known as nanogel-cross-linked (NanoClik) gels. The chemical-crosslinking points of self-assembled nanogels are first modified, and large-scale hydrogels are then prepared using these nanogels as building units. NanoClik gels can be created with different characteristics, including size, shape, and surface charge, and thus, the physical and mechanical characteristics of the resulting macro-sized hydrogel can be adjusted to suit various applications. Since the 2010s, various types of macro-sized NanoClik gels have been developed, known as nanogel tectonic engineering because nanogels are used as tecton units [51].

5.1 Nanogel-cross-linked gels

During the early stage of development, radical polymerization was used to create NanoClik hydrogels, and methacryloyl group-bearing CHP nanogels were used to provide effective cross-linking for gelation. The immobilized nanogels exhibit excellent chaperone-like activity for the refolding of chemically denatured proteins as well as the ability to capture and release protein [52]. Owing to the host–guest interactions of the cholesteryl group and cyclodextrin, the nanogels efficiently trap and release enzymes, which prevents the aggregation of heat-denatured carbonic anhydrase B [53]. In another research, CHP nanogels were modified with acrylate groups (CHPA) and cross-linked by poly(ethylene glycol) containing a thiol group (PEGSH) through the Michael addition reaction. The cross-linking reaction can be conducted under physiological conditions, such as at 37°C, and in the presence of phosphate-buffered saline. Following subcutaneous injection in mice, CHPA nanogels can provide sustained release of encapsulated interleukin-12 (IL-12) in the plasma [54]. By considering the results of various CHP nanogel modifications and cross-linker evaluations, CHP nanogels modified with acryloyl groups (CHPOA) and multi-armed PEGSH have been mostly used to create NanoClik gels. NanoClik gels containing protein-loaded nanogels can gradually release the nanogels by hydrolysis of CHP nanogels and PEGSH [55, 56]. In 2015, a porous NanoClik gel was created. According to two-photon excitation deep imaging, the pores are linked with widths in the range of several hundred micrometers. Owing to its constituent nanogels, the NanoClik porous gel can capture proteins through hydrophobic interactions, thus functioning as a chaperone. Moreover, in vivo evaluation indicated successful penetration of mouse embryonic fibroblasts into the network of interconnected pores [57]. Another study focused on the development of NanoClik microspheres for the manufacture of injectable sustained-release carriers. Self-assembled nanogels encapsulating proteins have been chemically cross-linked with biodegradable linkers to create NanoClik microspheres, which can release "drug-loaded nanogels" with a controlled drug distribution following sustained release, unlike conventional polymeric microspheres, which release naked drugs upon polymer dissolution [58]. In addition to CHP nanogels, cholesterol-bearing hydroxypropyl cellulose (CH-HPC) has been modified with acryloyl groups to prepare macro-scale NanoClik gels. HPC hydrogels are known as thermoresponsive building blocks because they can reversibly vary in size in water. Macrogels cross-linked with CH-HPC nanogels also exhibit this thermoresponsive activity [59].

5.2 Regenerative medicine

NanoClik gels are appealing prospects for a variety of regenerative medicine applications because they may be modified to have desired physical and mechanical properties as well as the capacity to release bioactive compounds in a controlled manner. Nanogel tectonic materials have properties that enable the gradual, regulated release of bioactive compounds, such as high porosity and water-swelling structure. This can be especially helpful for targeted drug delivery to specific tissue locations, and nanogel tectonic materials have been extensively investigated to build three-dimensional scaffolds that resemble the native tissue's extracellular matrix or tissue engineering.

In 2007, the effectiveness of a bone anabolic drug (prostaglandin E2, PGE2) delivered using a nanogel-based carrier was demonstrated. PGE2 is a hydrophobic small molecule, the encapsulation of PGE2 was conducted in the presence of ethanol. NanoClik gels consisting of CHPA and PEGSH in mice stimulated the creation of new bones without inducing weight loss. PGE2 alone in low doses could not stimulate the growth of new bone, whereas PGE2-loaded Nanoclik gels improved bone development [60]. This synthetic scaffold encapsulating bone morphogenetic proteins (BMP) activated osteoblasts and promoted bone growth for bone regeneration [61]. In a different study, a selective EP4 receptor agonist (EP4A) and low-dose BMP-2 were combined with this type of NanoClik gel. The combined material successfully activated bone cells and restored the outer and inner cortical plates as well as the bone marrow tissue of the calvarium [62]. In 2012, NanoClik gels consisting of CHPOA and PEGSH were used to deliver growth factors in a regulated manner over an extended period of time for bone tissue engineering. The material's ability to mend broken bones was assessed using BMP-2 and fibroblast growth factor 18, which indicated successful bone regeneration and osteoprogenitor cell infiltration [63]. As a regenerative therapy for tongue muscle abnormalities caused by surgical excision of tongue cancer, NanoCliP gels provided sustained delayed release of protein over the span of 14 days without an initial burst release. Mouse myoblasts attached to the NanoCliP hydrogel exhibited typical myotube development within the gel and remained viable for up to 7 days. Gel-encapsulated myoblasts transplanted into the tongue abnormality of a mouse model resulted in a notable rise in freshly formed myofibers [64]. In 2018, freeze-dried NanoClik porous (NanoCliP-FD) gel was developed to create three-dimensional scaffolds for bone tissue engineering. Fibroblast cells became directly converted osteoblasts (dOBs) by genetic reprograming. The NanoCliP-FD gel effectively supported fibroblast adhesion and formed a matrix of calcified bone after culturing. Animal studies showed that the NanoCliP-FD gel combined with cells reprogramed to dOBs considerably promoted bone regeneration in artificially produced bone defect lesions. These results suggest that the NanoCliP-FD gel is a promising treatment for bone disorders [65]. Subsequently, osteointegrative characteristics were investigated using the NanoCliP-FD gel as a scaffold for differentiation from mesenchymal cells to osteoblasts. Compared with a commercially available atelocollagen scaffold, the NanoCliP-FD gel, which contained chemically and physically cross-linked nanogels inside a porous gel, promoted the growth of apatite crystallites with an unusual c-plane orientation, reflecting the structure of natural enamel [66]. In 2022, researchers used in situ spectroscopic methods, particularly Raman spectroscopy, to show that mesenchymal stem cells cultivated on the NanoCliP-FD gel scaffold exhibited a greater rate of cartilage development and produced tissue [67].

A NanoClik membrane was created using CHPOA nanogels and PEGSH cross-linkers for wound healing. The NanoClik membrane considerably reduced the wound area and increased neoepithelialization. The NanoClik membrane also made it easier for collagen fibers to grow and assemble [68]. NanoCliP-FD sheets and fibers promoted the mending of significant bone lesions and were easy to transplant for bone regeneration. Compared with a control group of unconverted fibroblasts, dOBs treated with NanoCliP-FD in both sheet and fiber forms produced a much higher concentration of calcium deposits [69]. As another example, NanoClik microspheres (with an average diameter of 14 μm) were used to introduce functional support or spacers into cell spheroids. Mesenchymal stem cells from bone marrow were able to combine with NanoClik microspheres to form hybrid cell spheroids. NanoClik microsphere-based spacers were added without adversely affecting cell survival, showing that the microspheres served as a biocompatible scaffold for cell growth. Under culture conditions, the microspheres remained stable for two weeks. The capacity of the hybrid cell spheroids to be scaled up to the millimeter scale was also demonstrated, raising the possibility that they could be transplanted (**Table 1**) [70].

Application	Nanogels	Complexed with	References
Cancer vaccines	CHP	HER-2 protein and peptides	[27–30, 32]
	CHP	NY-ESO-1 protein	[31]
	CHP	mERK2, MAGE-A4 long peptide	[33, 37]
	CHP, CHPCOOH	Ovalbumin	[34–36]
Nasal vaccines	$CHPNH_2$	BoHc/A protein	[41]
	$CHPNH_2$	PspA protein	[42–45]
	$CHPNH_2$	ghrelin-PspA fusion protein	[46]
	$CHPNH_2$	NTHi surface antigen P6	[47]
Gene delivery	CH-CA-spe	siRNA	[48, 49]
	CH-CA-DEAE	CpG DNA	[50]
Regenerative medicine	CHPA	PGE2	[60]
	CHPA	BMP-2	[61, 62]
	CHPOA	BMP-2	[63]
	CHPOA	Cells	[64–67]

Table 1.
Medical application of nanogels consisting of cholesterol-bearing polysaccharides.

6. Conclusion

As described in this review, self-assembled nanogels consisting of cholesterol-bearing polysaccharides are one of the most well-studied biomaterials, and numerous medical applications have been reported in the past three decades. Although over 60 papers are covered, this review does not include all the medical applications of self-assembled nanogels, which have been used not only by researchers in academia but also by medical doctors and people in companies. Even now, there are not commercially available nanogels, while the practical use of the next generation of self-assembled nanogels is desired. In the near future, we hope to see innovative and optical applications of self-assembled nanogels.

Acknowledgements

We thank Edanz (https://jp.edanz.com/ac) for editing a draft of this manuscript.

Conflict of interest

Y. Tahara received funding from Asahi Kasei Corporation related to the research using hyaluronic acid-based nanogels.

Author details

Haddad Antonia-Nancy, Michiaki Matsumoto* and Yoshiro Tahara*
Department of Chemical Engineering and Materials Science, Doshisha University, Kyotanabe, Kyoto, Japan

*Address all correspondence to: mmatsumo@mail.doshisha.ac.jp; ytahara@mail.doshisha.ac.jp

References

[1] Kataoka K, Harada A, Nagasaki Y. Block copolymer micelles for drug delivery: Design, characterization and biological significance. Advanced Drug Delivery Reviews. 2001;**47**:113-131. DOI: 10.1016/S0169-409X(00)00124-1

[2] Akiyoshi K, Deguchi S, Moriguchi N, Yamaguchi S, Sunamoto J. Self-aggregates of hydrophobized polysaccharides in water. Formation and characteristics of nanoparticles. Macromolecules. 1993;**26**:3062-3068. DOI: 10.1021/ma00064a011

[3] Alemán JV, Chadwick AV, He J, Hess M, Horie K, Jones RG, et al. Definitions of terms relating to the structure and processing of sols, gels, networks, and inorganic-organic hybrid materials (IUPAC recommendations 2007). Pure and Applied Chemistry. 2007;**79**:1801-1829. DOI: 10.1351/pac200779101801

[4] Kabanov AV, Vinogradov SV. Nanogels as pharmaceutical carriers: Finite networks of infinite capabilities. Angewandte Chemie, International Edition. 2009;**48**:5418-5429. DOI: 10.1002/anie.200900441

[5] Menger FM. Supramolecular chemistry and self-assembly. Proceedings of the National Academy of Sciences. 2002;**99**:4818-4822. DOI: 10.1073/pnas.062524299

[6] Torchilin VP. Recent advances with liposomes as pharmaceutical carriers. Nature Reviews Drug Discovery. 2005;**4**:145-160. DOI: 10.1038/nrd1632

[7] Akiyoshi K, Deguchi S, Tajima H, Nishikawa T, Sunamoto J. Microscopic structure and Thermoresponsiveness of a hydrogel nanoparticle by self-assembly of a Hydrophobized polysaccharide. Macromolecules. 1997;**30**:857-861. DOI: 10.1021/ma960786e

[8] Akiyoshi K, Nishikawa T, Mitsui Y, Miyata T, Kodama M, Sunamoto J. Self-assembly of polymer amphiphiles: Thermodynamics of complexation between bovine serum albumin and self-aggregate of cholesterol-bearing pullulan. Polymer Journal. 1996;**112**:91-95. DOI: 10.1016/0927-7757(96)03560-1

[9] Akiyoshi K, Kobayashi S, Shichibe S, Mix D, Baudys M, Wan Kim S, et al. Self-assembled hydrogel nanoparticle of cholesterol-bearing pullulan as a carrier of protein drugs: Complexation and stabilization of insulin. Journal of Controlled Release. 1998;**54**:313-320. DOI: 10.1016/S0168-3659(98)00017-0

[10] Akiyoshi K, Sasaki Y, Sunamoto J. Molecular chaperone-like activity of hydrogel nanoparticles of Hydrophobized pullulan: Thermal stabilization with refolding of carbonic anhydrase B. Bioconjugate Chemistry. 1999;**10**:321-324. DOI: 10.1021/bc9801272

[11] Nomura Y, Ikeda M, Yamaguchi N, Aoyama Y, Akiyoshi K. Protein refolding assisted by self-assembled nanogels as novel artificial molecular chaperone. FEBS Letters. 2003;**553**:271-276. DOI: 10.1016/S0014-5793(03)01028-7

[12] Nomura Y, Sasaki Y, Takagi M, Narita T, Aoyama Y, Akiyoshi K. Thermoresponsive controlled Association of Protein with a dynamic Nanogel of Hydrophobized polysaccharide and Cyclodextrin: Heat shock protein-like activity of artificial molecular chaperone. Biomacromolecules. 2005;**6**:447-452. DOI: 10.1021/bm049501t

[13] Sawada S-I, Akiyoshi K. Nano-encapsulation of lipase by self-assembled Nanogels: Induction of high enzyme activity and thermal stabilization. Macromolecular Bioscience. 2010;**10**:353-358. DOI: 10.1002/mabi.200900304

[14] Sasaki Y, Nomura Y, Sawada S-I, Akiyoshi K. Polysaccharide nanogel–cyclodextrin system as an artificial chaperone for in vitro protein synthesis of green fluorescent protein. Polymer Journal. 2010;**42**:823-828. DOI: 10.1038/pj.2010.73

[15] Sasaki Y, Asayama W, Niwa T, Sawada S-I, Ueda T, Taguchi H, et al. Amphiphilic polysaccharide Nanogels as artificial chaperones in cell-free protein synthesis. Macromolecular Bioscience. 2011;**11**:814-820. DOI: 10.1002/mabi.201000457

[16] Ikeda K, Okada T, Sawada S-I, Akiyoshi K, Matsuzaki K. Inhibition of the formation of amyloid β-protein fibrils using biocompatible nanogels as artificial chaperones. FEBS Letters. 2006;**580**:6587-6595. DOI: 10.1016/j.febslet.2006.11.009

[17] Sasaki Y, Akiyoshi K. Nanogel engineering for new nanobiomaterials: From chaperoning engineering to biomedical applications. The Chemical Record. 2010;**10**:366-376. DOI: 10.1002/tcr.201000008

[18] Ozawa Y, Sawada S-I, Morimoto N, Akiyoshi K. Self-assembled Nanogel of Hydrophobized dendritic dextrin for protein delivery. Macromolecular Bioscience. 2009;**9**:694-701. DOI: 10.1002/mabi.200800288

[19] Takahashi H, Sawada S-I, Akiyoshi K. Amphiphilic polysaccharide nanoballs: A new building block for Nanogel biomedical engineering and artificial chaperones. ACS Nano. 2011;**5**:337-345. DOI: 10.1021/nn101447m

[20] Takeda S, Takahashi H, Sawada S-I, Sasaki Y, Akiyoshi K. Amphiphilic nanogel of enzymatically synthesized glycogen as an artificial molecular chaperone for effective protein refolding. RSC Advances. 2013;**3**:25716-25718. DOI: 10.1039/C3RA44572K

[21] Nakai T, Hirakura T, Sakurai Y, Shimoboji T, Ishigai M, Akiyoshi K. Injectable hydrogel for sustained protein release by salt-induced Association of Hyaluronic Acid Nanogel. Macromolecular Bioscience. 2012;**12**:475-483. DOI: 10.1002/mabi.201100352

[22] Antonia-Nancy H, Iwatsuki Y, Yabuuchi K, Aso S, Katsumata T, Fukumoto K, et al. Self-assembled nanogels based on hyaluronic acid for antibody protection from heat denaturation. Biochemical Engineering Journal. 2023;**196**:108955. DOI: 10.1016/j.bej.2023.108955

[23] Sawada S-i. Yukawa H, Takeda S, Sasaki Y, Akiyoshi K: Self-assembled nanogel of cholesterol-bearing xyloglucan as a drug delivery nanocarrier. Journal of Biomaterials Science. Polymer Edition. 2017;**28**:1183-1198. DOI: 10.1080/09205063.2017.1320827

[24] Kawasaki R, Sasaki Y, Katagiri K, Mukai S-A, Sawada S-I, Akiyoshi K. Magnetically guided protein transduction by hybrid Nanogel chaperones with iron oxide nanoparticles. Angewandte Chemie, International Edition. 2016;**55**:11377-11381. DOI: 10.1002/anie.201602577

[25] Kawasaki R, Sasaki Y, Nishimura T, Katagiri K, Morita K-I, Sekine Y, et al. Magnetically navigated protein transduction In vivo using iron oxide-Nanogel chaperone hybrid. Advanced

Healthcare Materials. 2021;**10**:2001988. DOI: 10.1002/adhm.202001988

[26] Tahara Y, Akiyoshi K. Current advances in self-assembled nanogel delivery systems for immunotherapy. Advanced Drug Delivery Reviews. 2015;**95**:65-76. DOI: 10.1016/j.addr.2015.10.004

[27] Gu X-G, Schmitt M, Hiasa A, Nagata Y, Ikeda H, Sasaki Y, et al. A novel Hydrophobized polysaccharide/Oncoprotein complex vaccine induces in vitro and in vivo cellular and humoral immune responses against HER2-expressing murine Sarcomas1. Cancer Research. 1998;**58**:3385-3390. DOI: none

[28] Shiku H, Wang L, Ikuta Y, Okugawa T, Schmitt M, Gu X, et al. Development of a cancer vaccine: Peptides, proteins, and DNA. Cancer Chemotherapy and Pharmacology. 2000;**46**:S77-S82. DOI: 10.1007/s002800000179

[29] Ikuta Y, Katayama N, Wang L, Okugawa T, Takahashi Y, Schmitt M, et al. Presentation of a major histocompatibility complex class 1–binding peptide by monocyte-derived dendritic cells incorporating hydrophobized polysaccharide–truncated HER2 protein complex: Implications for a polyvalent immuno-cell therapy. Blood. 2002;**99**:3717-3724. DOI: 10.1182/blood.V99.10.3717

[30] Kitano S, Kageyama S, Nagata Y, Miyahara Y, Hiasa A, Naota H, et al. HER2-specific T-cell immune responses in patients vaccinated with truncated HER2 protein complexed with Nanogels of cholesteryl pullulan. Clinical Cancer Research. 2006;**12**:7397-7405. DOI: 10.1158/1078-0432.CCR-06-1546

[31] Uenaka A, Wada H, Isobe M, Saika T, Tsuji K, Sato E, et al. T cell immunomonitoring and tumor responses in patients immunized with a complex of cholesterol-bearing hydrophobized pullulan (CHP) and NY-ESO-1 protein. Cancer Immunity. 2007;**7**:9. DOI: 10.1158/1424-9634.DCL-9.7.1

[32] Kageyama S, Kitano S, Hirayama M, Nagata Y, Imai H, Shiraishi T, et al. Humoral immune responses in patients vaccinated with 1-146 HER2 protein complexed with cholesteryl pullulan nanogel. Cancer Science. 2008;**99**:601-607. DOI: 10.1111/j.1349-7006.2007.00705.x

[33] Muraoka D, Harada N, Hayashi T, Tahara Y, Momose F, Sawada S-I, et al. Nanogel-based immunologically stealth vaccine targets macrophages in the medulla of lymph node and induces potent antitumor immunity. ACS Nano. 2014;**8**:9209-9218. DOI: 10.1021/nn502975r

[34] Miura R, Tahara Y, Sawada S-I, Sasaki Y, Akiyoshi K. Structural effects and lymphocyte activation properties of self-assembled polysaccharide nanogels for effective antigen delivery. Scientific Reports. 2018;**8**:16464. DOI: 10.1038/s41598-018-34885-8

[35] Miura R, Sawada S-I, Mukai S-A, Sasaki Y, Akiyoshi K. Antigen delivery to antigen-presenting cells for adaptive immune response by self-assembled anionic polysaccharide Nanogel vaccines. Biomacromolecules. 2020;**21**:621-629. DOI: 10.1021/acs.biomac.9b01351

[36] Miura R, Sawada S-I, Mukai S-A, Sasaki Y, Akiyoshi K. Synergistic anti-tumor efficacy by combination therapy of a self-assembled nanogel vaccine with an immune checkpoint anti-PD-1 antibody. RSC Advances. 2020;**10**:8074-8079. DOI: 10.1039/C9RA10066K

[37] Muraoka D, Seo N, Hayashi T, Tahara Y, Fujii K, Tawara I, et al. Antigen

delivery targeted to tumor-associated macrophages overcomes tumor immune resistance. The Journal of Clinical Investigation. 2019;**129**:1278-1294. DOI: 10.1172/JCI97642

[38] Hasegawa U, Nomura S-IM, Kaul SC, Hirano T, Akiyoshi K. Nanogel-quantum dot hybrid nanoparticles for live cell imaging. BBRC. 2005;**331**:917-921. DOI: 10.1016/j.bbrc.2005.03.228

[39] Ayame H, Morimoto N, Akiyoshi K. Self-assembled cationic Nanogels for intracellular protein delivery. Bioconjugate Chemistry. 2008;**19**:882-890. DOI: 10.1021/bc700422s

[40] Watanabe K, Tsuchiya Y, Kawaguchi Y, Sawada S-I, Ayame H, Akiyoshi K, et al. The use of cationic nanogels to deliver proteins to myeloma cells and primary T lymphocytes that poorly express heparan sulfate. Biomaterials. 2011;**32**:5900-5905. DOI: 10.1016/j.biomaterials.2011.04.058

[41] Nochi T, Yuki Y, Takahashi H, Sawada S-I, Mejima M, Kohda T, et al. Nanogel antigenic protein-delivery system for adjuvant-free intranasal vaccines. Nature Materials. 2010;**9**:572-578. DOI: 10.1038/nmat2784

[42] Kong Il G, Sato A, Yuki Y, Nochi T, Takahashi H, Sawada S, et al. Nanogel-based PspA intranasal vaccine prevents invasive disease and nasal colonization by Streptococcus pneumoniae. Infection and Immunity. 2013;**81**:1625-1634. DOI: 10.1128/iai.00240-13

[43] Fukuyama Y, Yuki Y, Katakai Y, Harada N, Takahashi H, Takeda S, et al. Nanogel-based pneumococcal surface protein a nasal vaccine induces microRNA-associated Th17 cell responses with neutralizing antibodies against Streptococcus pneumoniae in macaques. Mucosal Immunology. 2015;**8**:1144-1153. DOI: 10.1038/mi.2015.5

[44] Nakahashi-Ouchida R, Uchida Y, Yuki Y, Katakai Y, Yamanoue T, Ogawa H, et al. A nanogel-based trivalent PspA nasal vaccine protects macaques from intratracheal challenge with pneumococci. Vaccine. 2021;**39**:3353-3364. DOI: 10.1016/j.vaccine.2021.04.069

[45] Yuki Y, Uchida Y, Sawada S-I, Nakahashi-Ouchida R, Sugiura K, Mori H, et al. Characterization and specification of a trivalent protein-based pneumococcal vaccine formulation using an adjuvant-free Nanogel nasal delivery system. Molecular Pharmaceutics. 2021;**18**:1582-1592. DOI: 10.1021/acs.molpharmaceut.0c01003

[46] Azegami T, Yuki Y, Sawada S, Mejima M, Ishige K, Akiyoshi K, et al. Nanogel-based nasal ghrelin vaccine prevents obesity. Mucosal Immunology. 2017;**10**:1351-1360. DOI: 10.1038/mi.2016.137

[47] Nakahashi-Ouchida R, Mori H, Yuki Y, Umemoto S, Hirano T, Uchida Y, et al. Induction of mucosal IgA–mediated protective immunity against Nontypeable haemophilus influenzae infection by a cationic Nanogel–based P6 nasal vaccine. Frontiers in Immunology. 2022;**13**:819859. DOI: 10.3389/fimmu.2022.819859

[48] Toita S, Soma Y, Morimoto N, Akiyoshi K. Cycloamylose-based biomaterial: Nanogel of cholesterol-bearing cationic Cycloamylose for siRNA delivery. Chemistry Letters. 2009;**38**:1114-1115. DOI: 10.1246/cl.2009.1114

[49] Fujii H, Shin-Ya M, Takeda S, Hashimoto Y, Mukai S-A, Sawada S-I, et al. Cycloamylose-nanogel drug delivery system-mediated intratumor silencing

of the vascular endothelial growth factor regulates neovascularization in tumor microenvironment. Cancer Science. 2014;**105**:1616-1625. DOI: 10.1111/cas.12547

[50] Tahara Y, Yasuoka J, Sawada S, Sasaki Y, Akiyoshi K. Effective CpG DNA delivery using amphiphilic cycloamylose nanogels. Biomaterials Science. 2015;**3**:256-264. DOI: 10.1039/C4BM00293H

[51] Hashimoto Y, Mukai S-A, Sasaki Y, Akiyoshi K. Nanogel tectonics for tissue engineering: Protein delivery systems with Nanogel chaperones. Advanced Healthcare Materials. 2018;**7**:1800729. DOI: 10.1002/adhm.201800729

[52] Morimoto N, Endo T, Iwasaki Y, Akiyoshi K. Design of Hybrid Hydrogels with self-assembled Nanogels as cross-linkers: Interaction with proteins and chaperone-like activity. Biomacromolecules. 2005;**6**:1829-1834. DOI: 10.1021/bm050156x

[53] Morimoto N, Endo T, Ohtomi M, Iwasaki Y, Akiyoshi K. Hybrid Nanogels with physical and chemical cross-linking structures as Nanocarriers. Macromolecular Bioscience. 2005;**5**:710-716. DOI: 10.1002/mabi.200500051

[54] Hasegawa U, Sawada S-I, Shimizu T, Kishida T, Otsuji E, Mazda O, et al. Raspberry-like assembly of cross-linked nanogels for protein delivery. Journal of Controlled Release. 2009;**140**:312-317. DOI: 10.1016/j.jconrel.2009.06.025

[55] Shimoda A, Sawada S-I, Kano A, Maruyama A, Moquin A, Winnik FM, et al. Dual crosslinked hydrogel nanoparticles by nanogel bottom-up method for sustained-release delivery. Colloids and Surfaces, B: Biointerfaces. 2012;**99**:38-44. DOI: 10.1016/j.colsurfb.2011.09.025

[56] Shimoda A, Yamamoto Y, Sawada S-I, Akiyoshi K. Biodegradable nanogel-integrated hydrogels for sustained protein delivery. Macromolecular Research. 2012;**20**:266-270. DOI: 10.1007/s13233-012-0054-x

[57] Hashimoto Y, Mukai S-A, Sawada S-I, Sasaki Y, Akiyoshi K. Nanogel tectonic porous gel loading biologics, nanocarriers, and cells for advanced scaffold. Biomaterials. 2015;**37**:107-115. DOI: 10.1016/j.biomaterials.2014.10.045

[58] Tahara Y, Mukai S-A, Sawada S-I, Sasaki Y, Akiyoshi K. Nanocarrier-integrated microspheres: Nanogel tectonic engineering for advanced drug-delivery systems. Advanced Materials. 2015;**27**:5080-5088. DOI: 10.1002/adma.201501557

[59] Tahara Y, Sakiyama M, Takeda S, Nishimura T, Mukai S-A, Sawada S-I, et al. Self-assembled Nanogels of cholesterol-bearing Hydroxypropyl cellulose: A Thermoresponsive building block for Nanogel tectonic materials. Langmuir. 2016;**32**:12283-12289. DOI: 10.1021/acs.langmuir.6b02406

[60] Kato N, Hasegawa U, Morimoto N, Saita Y, Nakashima K, Ezura Y, et al. Nanogel-based delivery system enhances PGE2 effects on bone formation. Journal of Cellular Biochemistry. 2007;**101**:1063-1070. DOI: 10.1002/jcb.21160

[61] Hayashi C, Hasegawa U, Saita Y, Hemmi H, Hayata T, Nakashima K, et al. Osteoblastic bone formation is induced by using nanogel-crosslinking hydrogel as novel scaffold for bone growth factor. Journal of Cellular Physiology. 2009;**220**:1-7. DOI: 10.1002/jcp.21760

[62] Kamolratanakul P, Hayata T, Ezura Y, Kawamata A, Hayashi C, Yamamoto Y, et al. Nanogel-based scaffold delivery of prostaglandin E2 receptor–specific agonist in combination with a low

dose of growth factor heals critical-size bone defects in mice. Arthritis and Rheumatism. 2011;**63**:1021-1033. DOI: 10.1002/art.30151

[63] Fujioka-Kobayashi M, Ota MS, Shimoda A, Nakahama K-I, Akiyoshi K, Miyamoto Y, et al. Cholesteryl group- and acryloyl group-bearing pullulan nanogel to deliver BMP2 and FGF18 for bone tissue engineering. Biomaterials. 2012;**33**:7613-7620. DOI: 10.1016/j.biomaterials.2012.06.075

[64] Kinoshita N, Sasaki Y, Marukawa E, Hirose R, Sawada S-I, Harada H, et al. Crosslinked nanogel-based porous hydrogel as a functional scaffold for tongue muscle regeneration. Journal of Biomaterials Science. Polymer Edition. 2020;**31**:1254-1271. DOI: 10.1080/09205063.2020.1744246

[65] Sato Y, Yamamoto K, Horiguchi S, Tahara Y, Nakai K, Kotani S-I, et al. Nanogel tectonic porous 3D scaffold for direct reprogramming fibroblasts into osteoblasts and bone regeneration. Scientific Reports. 2018;**8**:15824. DOI: 10.1038/s41598-018-33892-z

[66] Horiguchi S, Adachi T, Rondinella A, Boschetto F, Marin E, Zhu W, et al. Osteogenic response of mesenchymal progenitor cells to natural polysaccharide nanogel and atelocollagen scaffolds: A spectroscopic study. Materials Science and Engineering: C. 2019;**99**:1325-1340. DOI: 10.1016/j.msec.2019.02.043

[67] Adachi T, Miyamoto N, Imamura H, Yamamoto T, Marin E, Zhu W, et al. Three-dimensional culture of cartilage tissue on Nanogel-cross-linked porous freeze-dried gel scaffold for regenerative cartilage therapy: A vibrational spectroscopy evaluation. International Journal of Molecular Sciences. 2022;**23**:8099. DOI: 10.3390/ijms23158099

[68] Maeda H, Kobayashi H, Miyahara T, Hashimoto Y, Akiyoshi K, Kasugai S. Effects of a polysaccharide nanogel-crosslinked membrane on wound healing. Journal of Biomedical Materials Research. 2017;**105**:544-550. DOI: 10.1002/jbm.b.33571

[69] Nakai K, Yamamoto K, Kishida T, Kotani S-i, Sato Y, Horiguchi S, et al. Osteogenic response to polysaccharide Nanogel sheets of human fibroblasts after conversion into functional osteoblasts by direct phenotypic cell reprogramming. Frontiers in Bioengineering and Biotechnology. 2021;**9**:713932. DOI: 10.3389/fbioe.2021.713932

[70] Hayashi S, Sasaki Y, Kubo H, Sawada S-I, Kinoshita N, Marukawa E, et al. Construction of hybrid cell spheroids using cell-sized cross-linked Nanogel microspheres as an artificial extracellular matrix. ACS Applied Bio Materials. 2021;**4**:7848-7855. DOI: 10.1021/acsabm.1c00796

Chapter 3

Polymeric Hydrogels and Nanogels: Classification, Development and Pharmaceutical Applications

Asif Mahmood, Hira Ijaz, Rai Muhammad Sarfraz, Nadiah Zafar, Muhammad Zaman and Mariya Azam

Abstract

This book chapter give an overview of natural and synthetic polymeric moieties consumed for developing hydrogels and their types. Different properties of nanogels are the advancement of hydrogels characterized by nano-size range, stimuli-responsive swelling, and release. Stimuli responsiveness is imparted by the presence of a suitable monomer. A number of polymerization approaches are presented in the literature that are employed to prepare such networks. These systems are elastic, rubbery, nontoxic, and biocompatible and offer prolonged release of the drugs without chances of dose dumping. These types of networks have potential pharmaceutical, agricultural, food, and biotechnological applications in terms of controlled, prolonged, and targeted drug delivery, solubility enhancements, stimuli-dependent intelligent drug delivery, such as contact lenses, wound healing, etc. In the current chapter, we have tried to introduce hydrogels and microgels, their different types, the variety of polymers used to develop such carrier systems, approaches to develop such drug delivery systems, and their utilization in various sectors in addition to the pharmaceutical sector.

Keywords: hydrogels, nanogels, free radical polymerization, drug delivery, wound healing

1. Introduction

Drug delivery encompasses a variety of techniques, formulations, and systems for delivering therapeutic moieties toward desired areas within the body to attain therapeutic outcomes effectively [1]. Conventional dosage forms are more likely required to be frequently administered in high doses, thereby resulting in certain side effects and patient noncompliance. To resolve such issues, controlled drug delivery systems with minimal side effects have been introduced that control the release of drugs in a number of ways [2]. By developing such delivery systems, the bioavailability of active agents can be potentiated through rate controlled delivery approach [3].

The term hydrogel has been used repeatedly in the literature. The first cross-linked polymeric network material was developed in 1960 [4]. It was equipped with properties of hydrogels, i.e., mainly having attraction toward water. Polyhydroxy ethyl methacrylate (PHEMA) hydrogel is massively employed in the preparation of contact lenses. These soft materials are also consumed by the patients for therapeutic purposes [5].

A new polymeric material was developed that was responsive to a number of environmental stimuli like pH, temperature, and concentration of different solutes in a solvent system. These stimuli play an important role in the swelling transition behavior of the networks and, subsequently, the release of therapeutic agents [6].

The third generation of hydrogels offered further advances in the field of polymeric materials. It has focused on stereo-complex materials, their interactions, and cross-linking through physical interactions [2]. Yana and his partners worked on collagen and shark cartilage tissues to develop a novel dressing that effectively decreased the healing burn [7]. Natural and synthetic hydrogels are used extensively in tissue engineering. Both of these hydrogels have variable potential applications in the control of thrombosis, drug delivery, biosensor coating, and cell transplant. Furthermore, the properties of hydrogels, like water-loving, biocompatibility, and their response to stimuli, established a great interest of researchers and scientists to introduce smart hydrogels [8].

Hydrogels are 3D water-loving carrier systems that have the ability to imbibe plenty of aqueous media or biological solvents without showing any transition in their structure [9]. These carrier systems, at their optimal swelling, show resemblance with body tissues because of their delicate and elastic consistency [10]. Hydrogels being intelligent carrier systems show several advantages like absence of active ingredient exposure in acidic environment, targeted drug delivery, and prolonged release of incorporated drugs [11]. A number of bonding types are involved development of chemically cross-linked hydrogels like ionic or covalent [5]. Presence of hydrophilic functional groups in developed polymer structure like –OH, –COOH, $–CONH_2$, –CONH, and SO_3H under suitable pH conditions offer absorption of a large amount of water by hydrogels and is helpful in the encapsulation of hydrophilic drugs [12]. Hydrogels are available in numerous forms, so classification may be based on various factors including method of preparation, source, biodegradation, response to certain stimuli, cross-linking, physical properties, etc. [13]. The detailed classification of hydrogels is summarized in **Figure 1**.

Hydrogels offer a variety of pharmaceutical, medical, and clinical applications like tissue regeneration, diagnostic applications, mucoadhesive potentials, etc. Hydrogels in literature are formulated in different shapes like nano and micro-sized particles, hydrogel membranes, beads, matrics, etc. [7–11], depending upon the required use.

2. Hydrogels

Hydrogels are high molecular weight network systems primarily consisting of hydrophilic polymers holding excess physiological fluid or water in polymer chains. Due to higher cross-linking reaction within polymers a grafted network structure is developed that prevent dissolution of hydrogels [14]. These type of network systems,

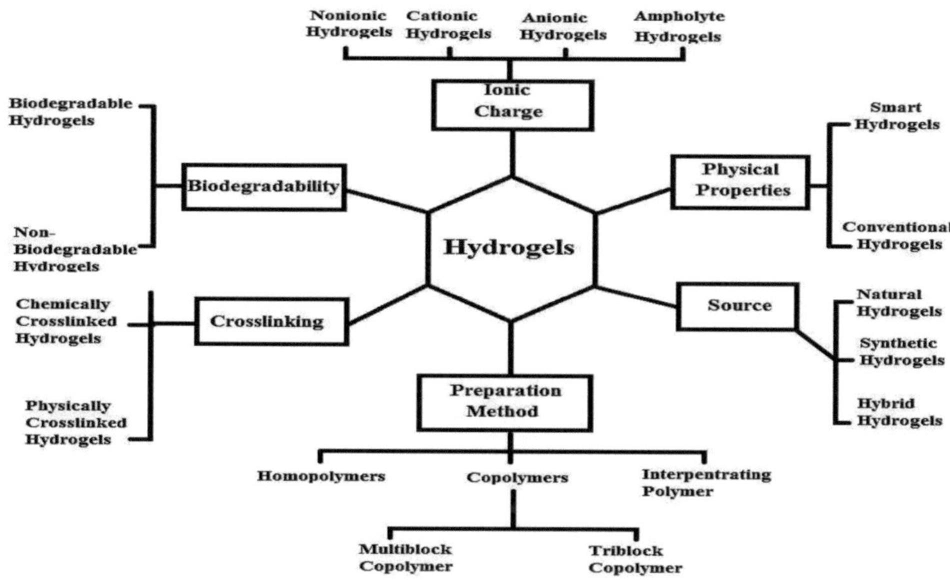

Figure 1.
Classification of polymeric hydrogels.

when fabricated to be influenced by certain external stimuli such as temperature, pH, enzyme, light, electric and magnetic fields, etc., are known as stimuli-responsive hydrogels [15]. On exposure to these environmental stimuli they exhibit response by undergoing volume or shape transitions [16, 17].

2.1 pH-responsive hydrogels

Hydrogels sensitive to pH are being developed extensively. There are various factors that control the swelling and de-swelling in hydrogels like ionic charge, extent of ionization, pka/pkb values, pH of medium, type of the monomer, concentration of polymer, and hydrophilicity of the developed networks [6]. At higher pH, weakly acidic functional groups present in anionic hydrogels like -COOH group undergo ionization resulting in swelling of the network. Similarly at lower pH, cationic hydrogels show swelling behavior due to ionization of basic functional groups like amines [18–20].

2.1.1 Mechanism of swelling in pH-responsive hydrogels

Swelling in hydrogels is caused by variations in osmotic pressure created within the network on exposure to physiological fluids that is controlled by the presence of charge on polymer, its hydrophilicity, and counter ions within the network system. Swelling of hydrogels proceeds in three steps [21]. Firstly, water penetrates into the hydrogel by diffusion, causing the hydration of polymeric chains. As a result, the relaxation of the polymeric chains promotes swelling in the grafted network [19]. Primary-bound water induces swelling due to its interaction with polar groups in the hydrogel, while secondary-bound water interacts with available hydrophobic groups.

An osmotic driving force is created that promotes diffusion of additional water called as free water into the hydrogel resulting them attaining equilibrium swelling. Flory and Rehner's theory describe this swelling response in terms of elasticity of polymeric chains and their compatibility with water molecules [22].

2.1.2 Mechanism of drug release in pH-sensitive Hydrogels

Active agents loaded or entrapped in hydrogels are released by different ways like diffusion-based, swelling-based, and chemically controlled approaches.

2.1.2.1 Diffusion-based controlled release

Cross-linked network structure of hydrogels composed of mesh-like geometry within the polymeric chains involved in the retention of small particles and liquids. The release of therapeutic agents through these hydrogels is described by size of mesh or pores [23]. Diffusion process is dominated in grafted network systems having mesh size greater than the size of drug [8]. In this case, size of pores has no significant impact on diffusion as smaller particles can pass easily through the cross-linked network. When the pores size in network becomes equal to the size of active moiety, then drug mainly diffuses through steric hindrance effect [21]. By using increased proportions of polymers or cross-linker contents in hydrogels pores size can be decreased. Due to reduced mesh size a greater frictional drag is exerted on drug by polymeric chains resulting in greater path length for drug transportation. In this case, diffusion is approximated by certain theoretical principles, and the net result is prolonged drug release by slow diffusion [24].

2.1.2.2 Swelling-based controlled release

Therapeutic agents can be released from hydrogels by another approach that is controlled by swelling mechanism [25]. Hydrogels swelling involves the absorption of physiological, biological, or buffer fluids that results in the enlargement of mesh size. Numerous factors can influence the swelling response like pH, temperature, light, ionic strength, electric field, glucose, etc. An important approach for cancer and oral drug delivery systems is pH mediated swelling [8]. By employing such technique certain pH responsive monomers and polymers are copolymerized to develop hydrogels that show negligible swelling in acidic pH of stomach, causing protection of entrapped drugs [26]. While at intestinal pH, these grafted networks show appreciable swelling attributing to drug release by swelling and diffusion modes. Numerous polymers having pendent basic or acidic functional groups are employed to develop such stimuli-responsive hydrogels. Furthermore, therapeutic moieties can be guided to targeted areas having solid tumors (more acidic environment) by using pH-responsive release systems (**Figure 2**) [27].

2.1.2.3 Chemically controlled release

Another strategy of drug release from hydrogels involves enzymatic or hydrolytic degradation of chains and follows a chemically controlled method [28]. By adopting this technique, release of drugs from cross-linked structures is kinetically controlled or

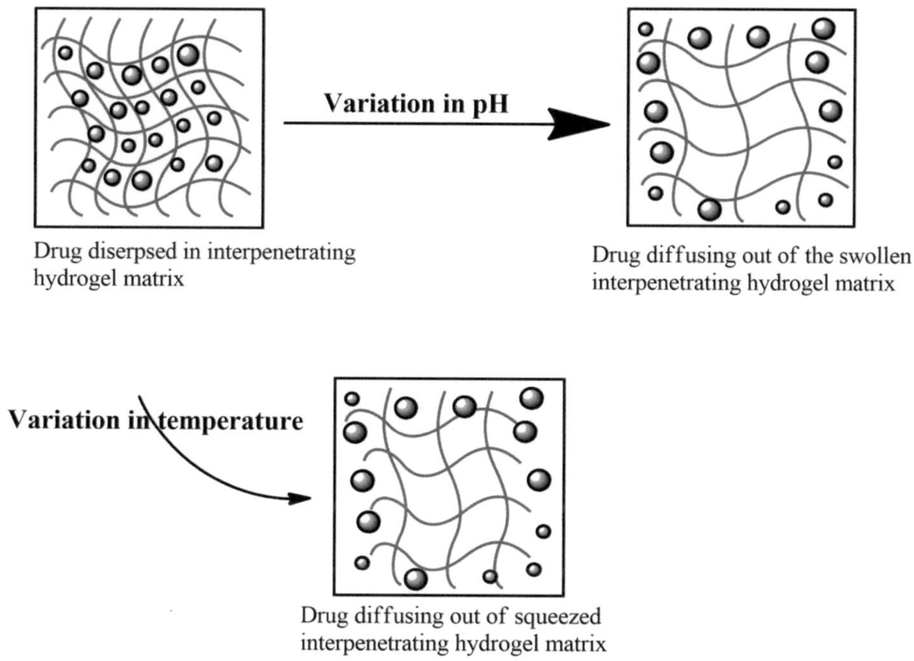

Figure 2.
Drug release mechanism in pH and temperature-responsive hydrogels.

reaction–diffusion controlled [29]. In kinetically controlled method, chemical reaction within hydrogel contents causes polymer decomposition by cleavage of its chains. In this case, there is no significant diffusion process involved. While, diffusion-controlled method results in drug release by both mechanisms involving polymer decomposition and followed by drug diffusion [30].

3. Properties of hydrogel

Hydrogels are declared biocompatible due to their high water content. These offer unique physical properties as their elasticity resembles the living tissues, due to which a lot of researchers have employed them for drug delivery applications. Structurally, these are highly porous, and their porosity can be tuned by the extent/degree of cross-linking in gel-matrix. Porosity is vital as it impacts on swelling, loading of drugs and release drug at desired rate depending upon the diffusion coefficient of the micro as well as macro molecules [28]. Hydrogels do not trigger or identified as foreign particles by body's own immune system due to the difference in interfacial tension between the surface of hydrogel and body fluids leading to minimized adsorption of protein and cell adhesion. Hydrogels are advantageous owing to ease of preparation. Hydrogels are capable of molding themselves according to the shape of surface applied [29]. Therefore, they are deformable. Nonirritant due to their soft and rubbery/elastic nature. Biodegradable hydrogels can be designed via implication of suitable hydrolytic, enzymatic, and environmental factors (e.g., pH, temp, or electric field).

4. Classification of hydrogels

These can be classified based on various criteria as follows (**Figure 3**) [31]:

1. Origin-based classification

   According to their source, hydrogels are categorized as:

   a. Natural

   b. Synthetic

   c. Semi-synthetic

2. Based on physical state

   a. Solid hydrogels

   b. Semi-solid hydrogels

   c. Liquid hydrogels

3. Polymeric composition-Based Classification

   a. Homopolymeric hydrogels

   b. Co-polymeric hydrogels

   c. Interpenetrating multipolymer hydrogels

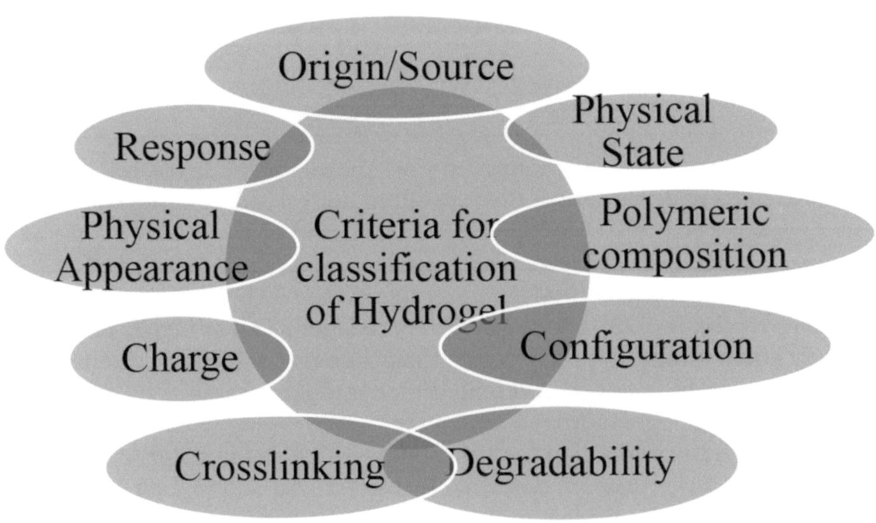

Figure 3.
Classification of hydrogels.

a. Homopolymeric hydrogels are formed by single monomer species, depending upon the nature of polymer and method of polymerization.

b. Co-polymeric hydrogels are made up of two or more monomer species. Among these, one is hydrophilic. They have arranged randomly, block-wise or alternatively along the polymeric chain network [32].

c. Interpenetrating multipolymer hydrogels are networks of two or more cross-linked polymers. One of which is natural, and the other is synthetic arranged in a network [33].

4. Configuration-based classification

On the basis of their physical and chemical configuration, there are four types:

a. Crystalline hydrogels

b. Semi-crystalline (complex amorphous-crystalline mixture) hydrogels

c. Amorphous hydrogels

d. Hydrocolloids hydrogels

5. Based on degradability

a. Biodegradable hydrogels

b. Non- biodegradable hydrogels

6. Cross-linking-based classification

Cross-linking may be chemical or physical depending upon the nature of cross-linked junctions; thus, hydrogels are classified into physically or chemically cross-linked hydrogels [34].

a. Chemically cross-linked hydrogels are made up of stable bonds.

b. Physically cross-linked hydrogels have unstable junctions which are formed either by polymer chain entanglements or as a result of physical interactions, hydrogen bonds, or hydrophobic interactions.

7. Based on physical appearance

a. Solid hydrogels

b. Semi-solid hydrogels

c. Liquid hydrogels

8. Charge-based classification

Cross-linked chains possess electric charges and classified accordingly.

a. Nonionic (neutral)

b. Amphoteric (composed of both acidic and basic groups)

c. Ionic (anionic/cationic)

d. Zwitterionic (having both anionic or cationic groups in each structural unit) [35]

9. Based on response to environmental stimuli

Hydrogels have the ability to respond to various chemical and physical stimuli [36, 37] and are classified into the followings:

a. Physically responsive hydrogels

b. Chemically responsive hydrogels

c. Biochemically responsive hydrogels

Physical stimuli may include temperature, light, electric or magnetic field, pressure, and sound. Chemical stimuli are pH, ionic strength, solvent composition, and molecular species. Biochemical stimuli include antigens, ligands, and enzymes [38].

5. Swelling behavior of hydrogel

Hydrogels get responsive to any kind of stimuli by means of volume alterations in their structure. In dry state, polymeric network of hydrogel is tightly bound and looks compacted, but as it comes in contact with fluid of suitable pH value, water molecules get diffused into its structure, and the polymer chains start separating, leading to subsequent swelling as a result of repulsion among ionized functional groups [31]. The extent of swelling depends upon three major factors:

I. Ionic constituents of hydrogels

Increase in the ionic content of hydrogels would result in higher electrostatic repulsion between ionized groups; thereby, elevation of swelling ratio occurs.

II. Cross-linking content of hydrogels

It controls the swelling of hydrogel system. Highly cross-linked network shows delayed swelling and vice versa.

III. Hydrophilic constituents

As hydrophilicity of hydrogel contents increases, there is more interaction between water and hydrophilic groups and ultimate increase in swelling [36].

Once swelling starts, it continues from surface to interior core of hydrogel matrix until equilibrium is achieved, which is indicated by optimum swelling. At peak swelling stage, polymeric chains maintain a greater distance, and the solvent absorbed creates a pressure on polymeric chains, which is compensated by the presence of a cross-linker [37, 39]. Through diffusion, solvent and solute gets in and out of the hydrogel structure. The theory for swelling of hydrogel, which narrates kinetic properties for transport of solvent molecule into hydrogel system, can be demonstrated as follows:

$$t = \frac{a^2}{D}$$

Where,
t = time required for swelling.
a = length of hydrogel.
D = diffusion coefficient

$$D = \frac{K + 4\mu/3}{f}$$

K = bulk modulus.
f = friction between polymer and bulk molecule.
μ = shear modulus.

Thus, the length of hydrogel has a direct relation with time of swelling thus the response of hydrogel to any stimuli is controlled by increasing or decreasing length of hydrogel. There are numerous strategies by which swelling of hydrogel can be enhanced, e.g., increasing the size of hydrogel by graft polymerization. This technique is beneficial for any material which de-swells promptly as the graft polymerization increases the sensitivity of hydrogel, and the resultant hydrogel responds much more rapidly by diffusion of solvent [40].

6. Limitations of hydrogels

Hydrogels being unique carrier systems have certain limitations along with different advantages.

i. Hydrogels offer poor tensile strength leading to early breakdown and hence dislocation from target sites.

ii. More significantly, drug delivery issues may arise, like heterogeneous drug loading, as in the case of hydrophobic drugs, which can affect drug release from the hydrogel network.

iii. Greater pore size promotes the uptake of larger water contents and hence faster drug release rates.

iv. Although hydrogels are deformable and can be used as injectable still, some require surgical implantation, which is exorbitant and demands sterile conditions for implantation [41, 42].

Constituents used in hydrogels are mentioned in **Table 1**.

Chemicals	Function
Okra gum	Main natural polymer
Cyclodextrin	Solubility enhancer
Acrylic acid (AA)	Monomer
Ammonium persulfate	Initiator
Methylene bisacrylamide	Cross linker
Capecitabine	Active ingredient

Table 1.
Constituents used in hydrogel.

7. Components needed for fabrication of IPNs

- Polymer: Main backbone of IPN.
- Monomer: Single units used in network fabrication by polymerization.
- Initiator: In order to kick-start the polymerization reaction.
- Cross-linking agents: Cross-links polymeric chains for network fabrication to resist disintegration and dissolution of IPN itself.

7.1 Drug loading and drug release

- Drug loading in IPNs:

  Generally, two methods are followed for drug loading in IPNs; preloading and postloading [43].

- Preloading: Drug loading in this method is carried out during the preparation of polymeric networks. Preweighed amount of medication (active pharmaceutical ingredient) is solubilized in the most suitable solvent, and afterward, this drug solution is added to the final polymerization mixture or solution. Another approach is to blend medication solution with a monomer solution, and other solutions are added separately. After the completion of polymerization, API will remain entrapped within the network framework.

- Postloading: This method involves the formation 1% solution of medication. This is done by dissolving 1 g of medication (drug) in the most suitable solvent, and then the volume is made up to 100 ml. Dried and preweighed hydrogel discs are dipped in drug solution for predefined intervals until and unless maximum swelling occurs. Loaded IPNs are dried at normal room temperature. Sometimes, drying is accomplished by using an oven for a particular time period and temperature. In this way, dried and drug-loaded IPNs are obtained.

- Drug release in IPNs:

  Generally, three methods are being followed for drug release in IPNs [37];

7.2 Swelling- controlled release

When water or biological fluids comes in contact with the dried or dehydrated IPNs it imbibes into the voids in between the polymeric chains. Thus, the drug, after its dissolution, is released in solution form from the networks.

7.3 Diffusion-controlled release

Drug release from IPNs is done by following the diffusion law (Fick's diffusion law), i.e., the drug is released according to the concentration gradient [23].

7.4 Chemically controlled release

Chemical reactions like enzymatic reaction and hydrolysis cause cleavage of polymeric chains resulting in drug release. Drug release from IPNs is studied by performing dissolution studies, and then it is validated through different kinetic models like zero-order, first order, Higuchi, Korsmeyer–Peppas, and Hixson Crowell models to find out the drug release kinetics in terms of best-fit model and mechanism of drug release from regression coefficient and value of exponent n, respectively [39].

8. Polymers used for hydrogels

Numerous polymers from natural and synthetic sources are being used for development of hydrogels. Polymers obtained from natural sources have excellent characteristics like biocompatibility, biodegradability, nontoxic nature, and economical, but hydrogels developed by these polymers show poor durability [40, 43]. While synthetic polymers are expensive and nonbiodegradable, and their hydrogels are good in mechanical strength. So these limitations can be overcome by blending the natural polymers with synthetic ones to develop more stable hydrogel networks [41, 42]. Natural polymers can be categorized as plant-based and animal-based polymers that have been extensively employed for various applications. Gums obtained from plants have gained considerable interest in drug release applications owing to their biocompatibility, low cost, and easy accessibility [44]. Detail of some natural and synthetic polymers that are being employed for hydrogel network formation given in the proceeding sections.

8.1 Pectin

Pectin is a carbohydrate-based polymer having a natural origin. It is prepared by extraction process on a commercial scale; from different types of citrus plants like apples, guavas, strawberries, oranges, and grapes under slightly acidic conditions [44]. High methoxyl pectin and low methoxyl pectin are two groups in which pectin can be divided. In the food industry, it is employed as a gelling agent as well as a stabilizer. Pectin has already been investigated for the development of targeted delivery of drugs and other biomedical applications [45]. Rapid degradation of pectin occurs by colonic microorganisms and thus making it a possible carrier for colon-targeted drug delivery because of its exceptional biocompatible and biodegradable

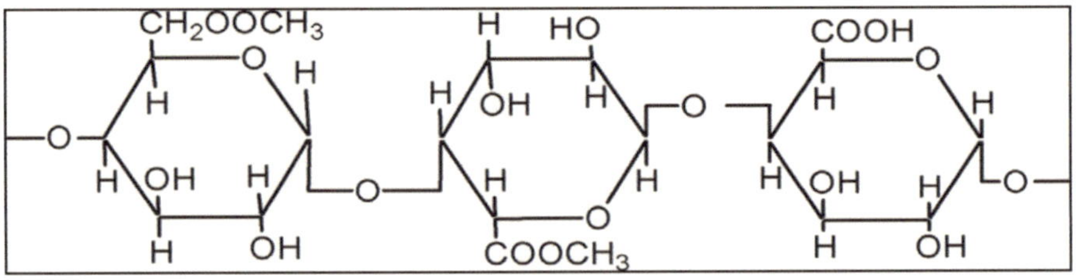

Figure 4.
Chemical structure of pectin.

nature [46]. Nowadays, scientists have been working on the fabrication of pectin-based preparations such as polymeric hydrogels, tablets, microspheres, films, and nanoparticles [47].

Structurally, pectin is a polysaccharide. It consists of repetitive galacturonic acid units joined together through α-1,4-glucoside bonds, as shown in **Figure 4** [48]. Pectin has linear as well as a branched structure, thus making it a complex molecule. Methylation and esterification both contribute to the gelling ability of pectin. The degree of esterification not only contributes to the gelling potential of pectin but also affects solubility profile of pectin. Pectin forms aggregates in cold water but solubilizes in slightly warm water under continuous stirring [49].

8.2 Natrosol or hydroxyethyl cellulose (HEC)

Natrosol is a polymer that is derived from cellulose. It is a water-soluble and nonionic derivative of cellulose. The basic structure is shown in **Figure 5**. It has a chain structure in which anhydrous glucose units are attached to each other by β-1,4-linkage giving rise to three free hydroxyl groups in each unit. These free hydroxyl groups make it a suitable polymer for hydrogel fabrication [50].

It is marketed in two grades; R-grade and B-grade. The R-grade of Natrosol facilitates preparation of solution in water without formation of lumps. Natrosol has been used in various pharmaceutical preparations because of its emulsifying, water-retaining, thickening, gelling, and stabilizing properties. It is most widely consumed in ophthalmic and topical preparations to facilitate drug delivery. It is also used in household products like lubricants, detergents, and cosmetics attributed to its water-soluble and nonionic nature [51]. It is soluble in hot water as well as cold water. Its viscous nature has indirect relation with increasing temperature can be reduced by increasing temperature. This property also makes it a useful ingredient for making gels at lower temperatures [52].

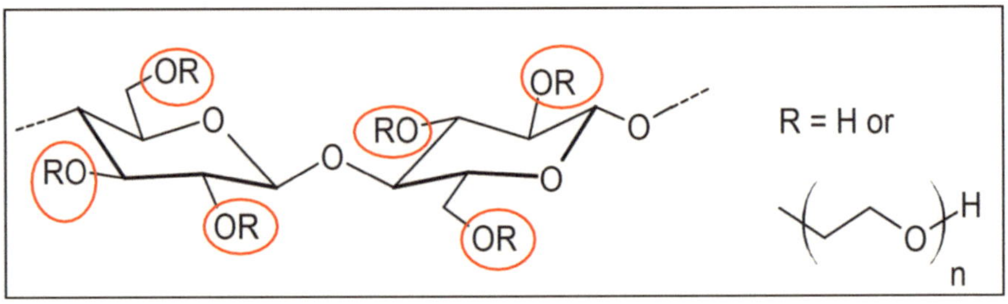

Figure 5.
Chemical structure of natrosol.

8.3 Tamarind gum

Tamarind gum is a polysaccharide, a natural polymer obtained by extraction of *Tamarindus indica* seeds. Its structure consists of β-(1,4)-D-glucan attached with pendant chains of α-(1,4)-D-xylopyranose and [β-D-galactopyranosyl-(1,2)-α-D-xylopyranosyl] linked to glucose units [53]. It is extracted by employing enzymatic and chemical-based extraction methods. In chemical method, ground seeds are soaked in water to prepare a slurry, which is then added to boiling distilled water with continuous stirring. The resultant mixture is filtered and poured in twice its volume of ethanol to get the required precipitates [54]. These precipitates are dried to prepare the gum. In enzymatic-based extraction, the protease enzyme is allowed to act on tamarind seed powder after mixing it with ethanol. Mixture is then centrifuged, and the supernatant collected is poured into ethanol to obtain precipitates of tamarind gum. These are dried after separation for further use [55].

It contains abundant hydroxyl groups. It is transformed into mucilage while contacting with water. This can be used in the pharmaceutical sector as an excipient because of its compatibility with living tissues, nontoxic nature, and bioadhesive potentials (**Figure 6**) [56].

8.4 Fenugreek

Fenugreek seeds gum is obtained by extraction of seeds endosperm of fenugreek (*Trigonella foenum-graecum* L.) [57]. High content of gum is present in fenugreek seeds that are converted into thick and viscous mass on exposure to water and biological fluids. Main constituents of gum are galactomannans which are polysaccharides in which single D-galactose units are linked to main β-(1–4)-D-mannan backbone through α-(1-6)-glycosides bond [58, 59]. The gum is extracted by soaking the fenugreek seeds in distilled water (1:10 w/v) at 40°C for 4 hours with occasional shaking. After that thick mixture is filtered through muslin cloth, and ethanol is added to the filtrate (1:1 v/v). Resulting precipitates of gum are kept in oven at 40–45°C till drying. Fenugreek seed gum has numerous applications in food industry and pharmaceutical formulations as binding, suspending, gelling agent, etc. (**Figure 7**) [60].

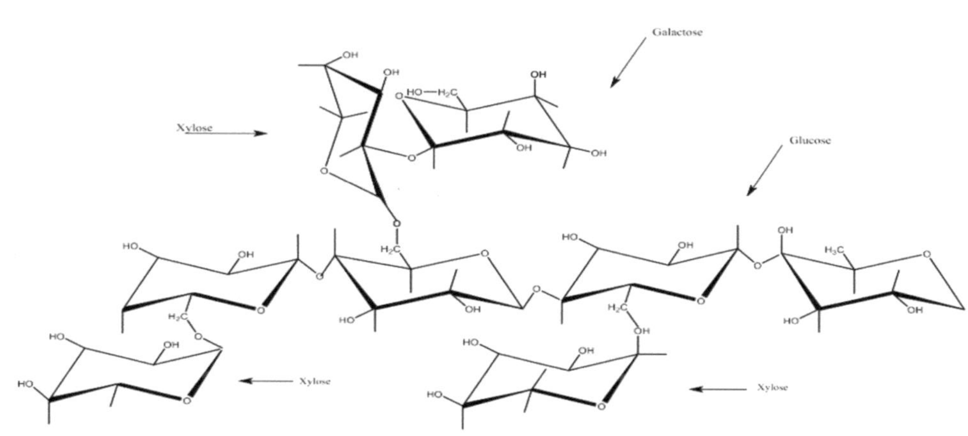

Figure 6.
Structural formula of tamarind seed polysaccharide.

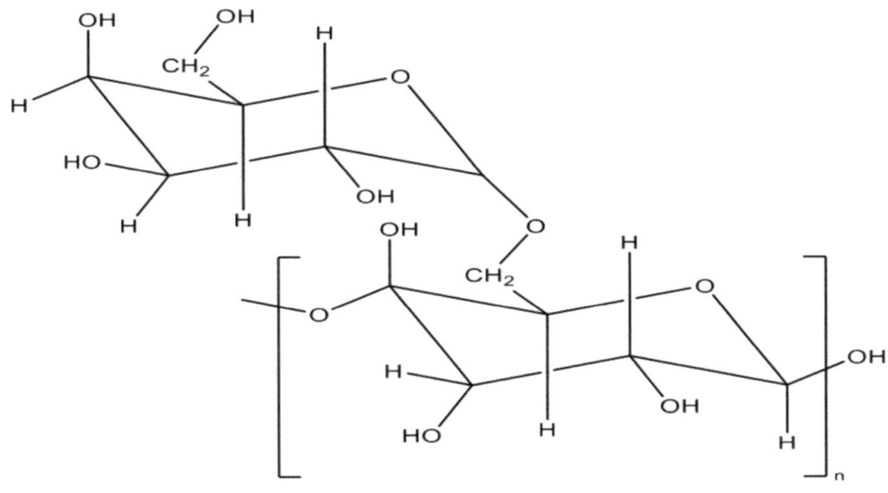

Figure 7.
Structural formula of fenugreek seed gum.

8.5 Chitosan

Chitosan is commercially developed by deacetylation of chitin and is mainly comprised of D-glucosamine and *N*-acetyl-D-glucosamine units having β-(1-4)-linkages [61]. It is available as high, medium, and low molecular-weight chitosan. This polymer is cationic, contains nitrogen in its chemical structure, and has ability to form polyelectrolyte complexes [62]. It becomes water soluble by forming carboxylate salts like acetate, citrate, formate, malate, glycolate, pyruvate, lactate, and ascorbate. These properties make it different from other polysaccharides [63]. Chitosan is widely employed in biotechnology and biomedical areas owing to its stable, nontoxic, and biodegradable characteristics [64]. Various controlled-release formulations have been developed by chitosan keeping in view its ideal properties. Furthermore, it is largely employed for the development of hydrogels in different forms including microparticles, microspheres, beads, sponges, composites, tablets, etc. (**Figure 8**) [65].

8.6 Agarose

Agarose is a neutral polysaccharide found in certain seaweeds [66]. Its chemical structure consists of agarbiose units in which 3,6-anhydro-α-L-galactopyranose are linked to C-1 and C-3 positions of β-D-galactopyranose. Agarose dissolves in boiling water and gels on lowering the temperature (30–40°C) [67]. The gelation mechanism of this polymer involves self-assembly of its units along with molecular cross-linking

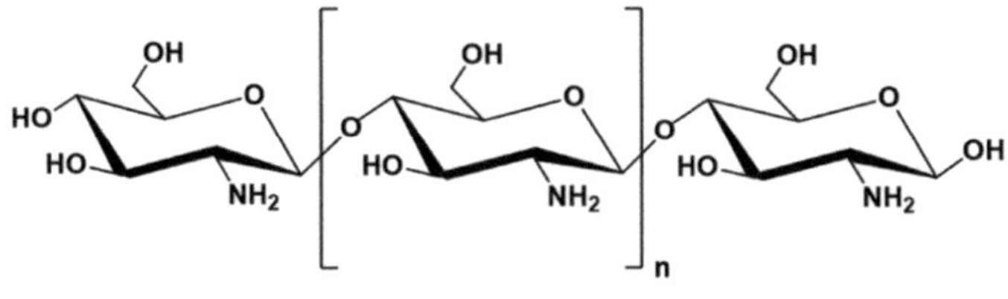

Figure 8.
Chemical structure of chitosan.

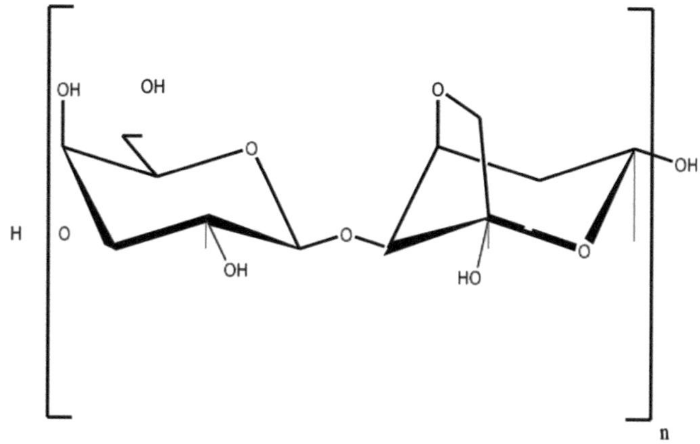

Figure 9.
Structural formula of agarose.

and conformational change [68]. Different factors affecting this mechanism are rate of agitation, temperature, concentration of polymer, co-solutes, and solvent properties [69]. Mechanism of gel formation is not fully understood; however, it can be explained regarding liquid–liquid phase separation. Hydrogels developed by using pure agarose are transparent, hard, elastic, and thermoreversible (**Figure 9**) [51].

8.7 Cyclodextrins

Cyclodextrins (CDs) are very familiar oligosaccharides containing 6 to 7 glucose units that are linked together via α-1, 4 glucosidic linkages. Cyclodextrins are stable at lower pH regions like stomach and intestine, but these are fermented at higher pH regions, i.e., colon. These have nest, like geometry with inner hydrophobic lining and outer hydrophilic end. Microflora of colon is capable of degrading and completely digest within the colon; hence these polymers can be employed for drugs to be deliver in colon. The structural formula is shown in **Figure 10**.

8.8 Polyethylene Glycol 8000

Polyethylene Glycol 8000 is a polymer that can be prepared by combining ethylene oxide with water. PEG occurs as white or off-white flakes that are waxy in nature with a faint and sweet odor [70]. This form is a result of condensation reaction between ethylene oxide and water under pressure in the presence of a catalyst. It is a highly water-soluble polymer that normally does not cause any harm to skin and is also stable in nature [71]. They are capable of promoting dissolution and systemic availability of numerous less water-soluble drugs. Different grades of PEG are generally used in wide range of pharmaceutical formulations (**Figure 11**) [72].

8.9 Okra gum

Okra is one of the traditional plant and a multipurpose crop with botanical name *Hibiscus esculentus* belongs to the family Malvaceae. It is an annual plant cultivated in tropical and subtropical areas of the world [73]. It can grow on all soil types and is the most tolerant plant. It is superfluous in nutritional content. Different parts of the plant

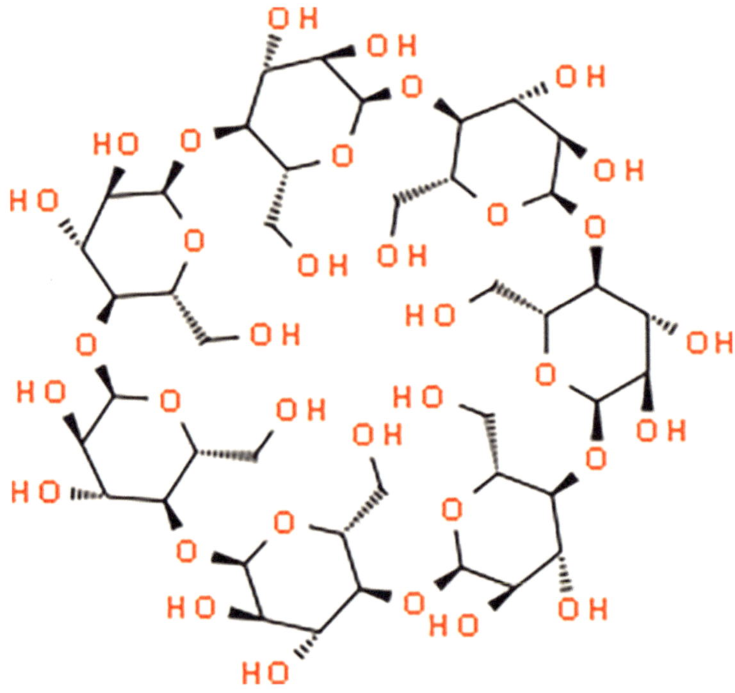

Figure 10.
Structural representation of cyclodextrin.

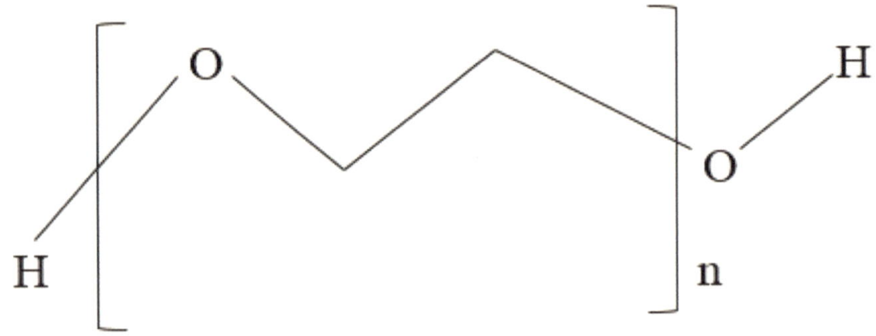

Figure 11.
Structural formula of polyethylene glycol.

are extensively used in traditional medicines as antidiabetic, diuretic, and anticancer agents [74]. It has wide applications in the pharmaceutical and food industries. Okra is abundant in bioactive compounds like flavonoids, polysaccharides, polyphenols, caffeine, and pectin. The whole plant is edible and is used for food, nonfood, and medical applications.

As compared to synthetic polymers, natural polymers offer good biocompatibility and biodegradability. However, natural polymers yield less mechanical strength in polymeric formulations [75]. Okra gum is chemically inert, nonirritant, biodegradable, biocompatible, and eco-friendly. It is economical and widely cultivated crop. It is comprised of L-rhamnose, D-galactose, and L-galacturonic acid in its composition (**Figure 12**) [76].

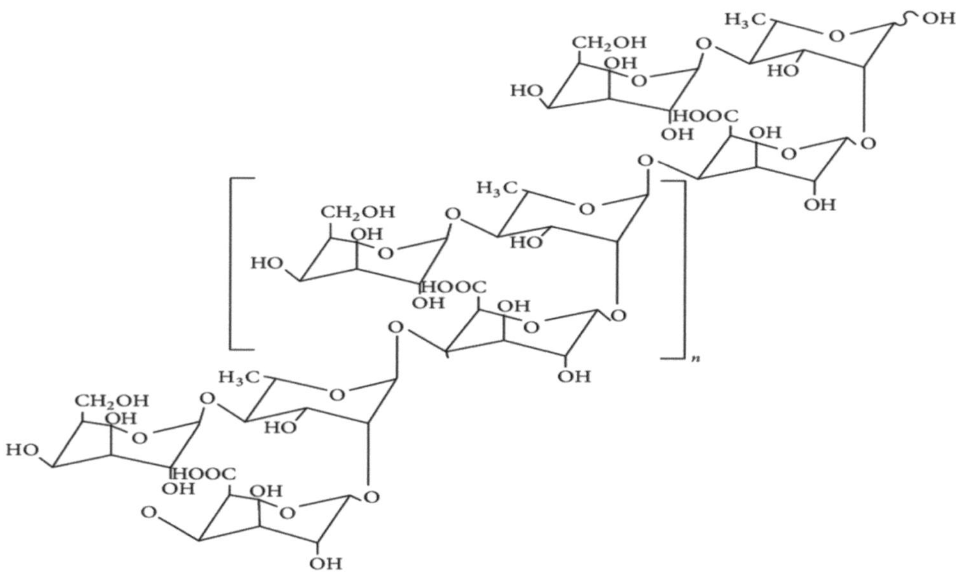

Figure 12.
Chemical structure of okra gum polysaccharide.

In pharmaceutical industry, although its role has not been explored enough as hydrogel preparations, it is evaluated as an economical and effective binding agent in solid dosage forms [45, 47, 48], in forming controlled release matrices [49, 50], for preparation of sustained release beads [51], microspheres [52, 53], in mucoadhesive nasal gel [54] and in many other formulations as a sustained release polymer.

8.10 Xanthan gum

It is obtained from sucrose or glucose by the process of fermentation by *Xanthomonas campestric*. It is soluble in both hot and cold water. Xanthan gum has stability over a pH ranging from 4 to 10. It is used in cosmetic, topical formulation and as a stabilizer in food industry [77].

8.11 Hyaluronic acid

It is also known as hyaluronate and hyaluronan. Hyaluronic acid is present in the tissues of animals. It is composed of N-acetyl-D-glucosamine and D-glucuronic acid, which are bonded together by β-(1-3) link [78]. It is used for ophthalmic, parenteral, and nasal preparation owing to its nonimmunogenicity and biocompatibility [79].

8.12 Gelatin

It is a natural polymer that has aqueous solubility. Collagen is hydrolyzed to obtain gelatin, where collagen is present in the bone and connective tissues of animals. Gelatin is insoluble in alcohol, whereas it is soluble in a hot mixture of water and glycerin. It has wide applications in protein and drug delivery, cell culture, and gene delivery [80, 81].

8.13 Polyethylene glycol

It has aqueous solubility. Hydrogels based on PEG are also known as intelligent or smart hydrogels due to their stimuli dependence. The stimuli can be either chemical (specific ion or pH) or physical (radiation, light, pressure, solvent, or temperature). PEG base hydrogels are widely used for the controlled release of therapeutic agents [82, 83].

8.14 Polyvinyl alcohol

It is an aqueous soluble polymer. Hydrogels based on polyvinyl alcohol are prepared by freeze and thawing process. Their water retention capacity is high; therefore, they are used in wound dressing. PVA is also used in cartilage reconstitution and contact lenses [84, 85].

8.15 Polyvinyl pyrolidone

It has high solubility in polar solvents as well as in water. It has good elasticity, and the production cost is less. PVP is combined with other different polymers such as CMC to enhance biocompatibility and mechanical properties [86, 87].

8.16 Pluronic acid F127

Pluronic acid F127 is composed of two blocks, a water-loving polyethylene oxide (PEO) and water-hating polypropylene oxide (PPO). These polymer blocks are arranged in a basic tri-block structure of repeating units, i.e., A-B-A. Molecular weight of F127 is approximately 12,600 Da. At temperature above lower critical solution temperature, i.e., 30°C, F127 loses solubility of its PPO block, leading to micelle development, and solution is converted into gel [88]. Pluronic acid serves as an important base for thermo-sensitive hydrogels. For localized malignancy treatment, intra-tumoral or peri-tumoreal infusion of gel formed by pluronic acid F127 leads to the development of 'depot' at the point of administration that gradually and constantly releases medicinal agents to the malignant and neighboring tissues [89]. Utilizing a topical or injectable gel for physical targeting has added preferences over passive or other actively targeted treatments that possess ability to supply medicinal agent to the whole tumor irrespective of vascular status, thus giving precise dosing of the drug without causing systemic toxicity [90]. Pluronic Acid F127 has a very vital job among in situ gelling frameworks due to its sol–gel transition character, a solution of poloxamer in water at a low temperature converts into polymeric gel at a higher temperature. Due to this capability, PF127 has confirmed its use in medical and therapeutic fields like in fertility management cases, pain management, and delivery of drugs at a controlled rate via eyes or rectal route (**Figure 13**) [91].

IUPAC name: Poly (ethylene glycol)-block-poly (propylene glycol)-block-poly. (ethylene glycol)

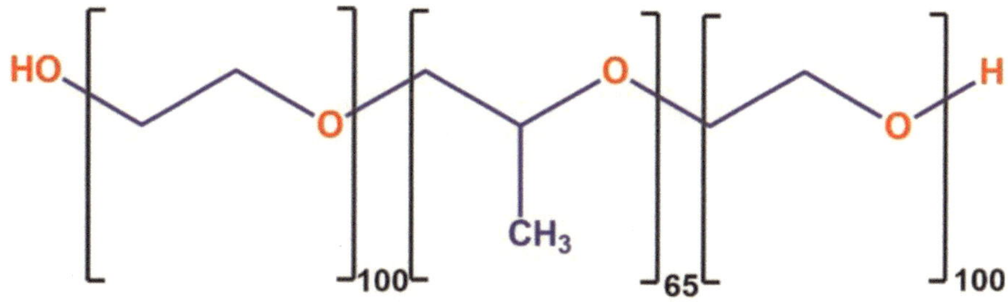

Figure 13.
Structure of pluronic acid F127.

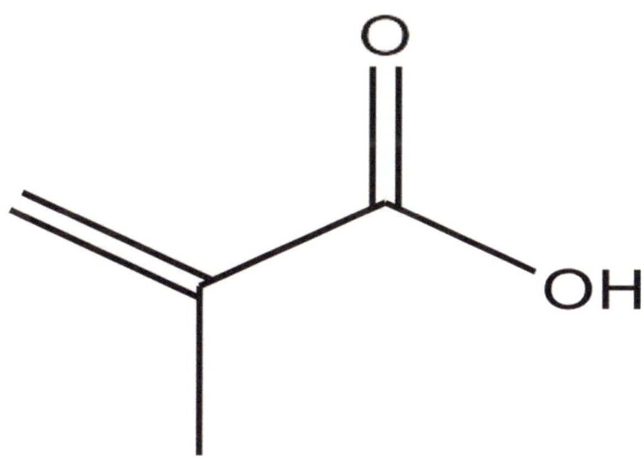

Figure 14.
Structural formula of MAA.

9. Other excipients for hydrogels

9.1 Methacrylic acid

Methacrylic acid (MAA) is a colorless organic liquid and has a strong pungent odor. It is widely employed as a monomer due to its excellent pH-responsive characteristics [92]. Hydrogels developed from this monomer exhibit appreciable swelling in basic media, while this effect is negligible in acidic media. This effect is contributed by ionizable carboxylic groups from MAA structure, creating repulsion within the network (**Figure 14**) [93].

9.2 Acrylic acid

It is an organic compound and a simplest form of unsaturated carboxylic acid. It also contains a vinyl group, which is directly attached to a carboxylic acid. It is a colorless liquid with a characteristic acrid odor and a tart smell. It is miscible with aqueous media and weakly acidic compounds. It is highly corrosive and irritant to skin. It is also known as 2-propenoic acid. In literature, it is widely used as an initiator in various hydrogel preparations (**Figure 15**) [94, 95].

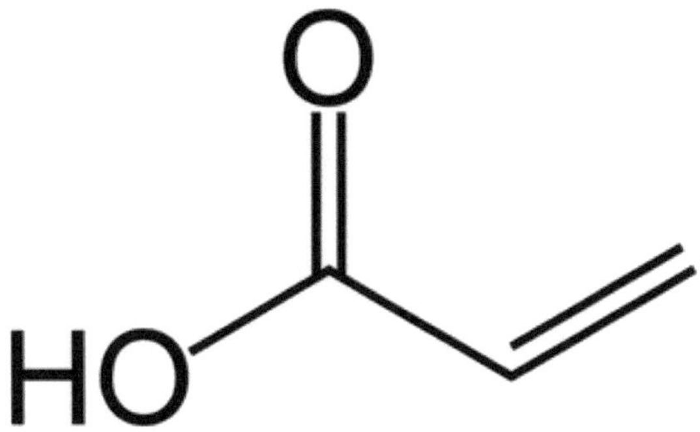

Figure 15.
Structural representation of AA.

9.3 *N,N′*-methylenebisacrylamide

A bi-functional monomer, *N,N′*-methylenebisacrylamide (MBA) consists of two double bonds in its structure. It is majorly used in polymerization reactions due to its cross-linking properties [44]. It is a hydrophilic, cross-linking agent used during polymerization reactions. It has the ability to form a network structure instead of making linear chains, thus assisting in maintaining the firmness of prepared gel. It has diverse applications as a monomer and cross-linker in various polymerization reactions. Chemically, it contains two unsaturated bonds and have been employed in the preparation of various pharmaceutical dosage forms (**Figure 16**) [96].

10. Different methods of hydrogel preparation

A number of methods have been quoted in literature to prepare polymeric networks. These include ionic gelation, spray drying, dispersion photopolymerization, suspension cross-linking technique, ionotropic gelation method, free radical

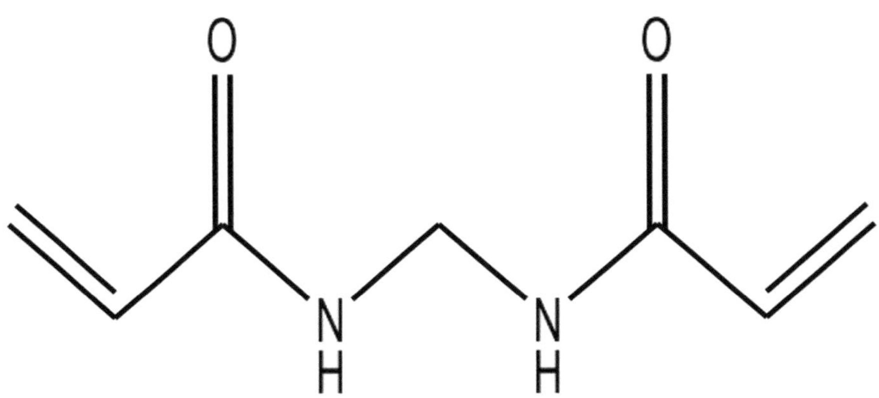

Figure 16.
Structural formula of MBA.

polymerization, membrane emulsification technique, free radical precipitation polymerization, Michael addition reaction, and inverse emulsion polymerization techniques depending upon the objective of the study.

11. Applications

11.1 Biomedical applications

The hydrogels exculpate oneself comparably to anthropoid organs in reply to climate state such as change in pH, change in temperature, electric field, and enzymes. They are cast off in pharmaceuticals lodge, diagnostic gadgets, bone implants, embolus, and computerized grasp [17, 46, 97]. They are also employed in urinary catheter by intercepting them from germs attack on the surface [98]. Kampa [99] successfully used the hydrogel electrodes in the diagnosis and treatment of body tissue [18].

11.2 Biotechnology applications

Hydrogels are also giving an important role in various mechanisms. They are used in the refining of various gadgets and centralizing the moist forum [19]. Murphy et al. [100] have worked on the use of hydrogels in the treatment of wounds. They have worked on the development of new technology containing hydrogel-based bioprinting. This technology was used for faster healing of wounds. They find out that hyaluronic acid-based hydrogels are best for bio printing applications [101, 102].

11.3 Pharmaceutical applications

There are various petitions for hydrogels in pharmaceutical factories. They are mostly used for the intended site distribution of drugs, sustained delivery of drugs, controlled release of medicine, and reducing the frequency of drugs [21]. Peppas et al. have studied hydrogels to find a suitable way to cure a number of diseases. This study showed that hydrogels are water-loving and can store bulk amounts of water and medicaments. These medicaments can be made to release as desired in response to different stimuli. The release of drugs from hydrogel can be controlled [103].

11.4 Separation technology

Currently water has been polluted broadly by the factories' consumption containing various categories of pigments from textile factories, pharmaceuticals, nonmedical, and paper industries. Hydrogels play a vital character in the separation of dyes by acting as adsorbents [22, 23].

11.5 Agriculture industry

In the cultivation part, hydrogels give a dominant part by assuming control availability of supplements to crops and plants from loaded hydrogels [25]. The advancement of a cellulose-based biodegradable superabsorbent hydrogel has been studied for the streamlining of water assets in agriculture. The sorption capacity of the proposed hydrogel was rightly evaluated, with explicit respect to two factors that may assume a key job in the soil condition: ionic strength and pH. A fundamental assessment of the hydrogel's potential to store water was performed by utilizing the hydrogel in greenhouses to develop tomatoes. The soil and water retention curve was noted. The results demonstrated that the material permitted productive storage and spontaneous supply of water to the soil and the plant roots [26].

11.6 Contact lenses

Hydrogels are used to create contact lenses that envelop the enormous mass of the essential ingredients needed to be used in different situations. There are some requirements to make it safe during use for a hydrogel material that is used as a contact trifocal. Such essential elements contain amounts of water, oxygen pores, and great focus point. So as to construct the water tranquil of hydrogel and to acquire additional bulging effects, different kinds of monomers can be used. These include dihydroxy methacrylates, acrylamides, and MAA [27, 28].

11.7 Food packaging

Some analysts are working on hydrogels as a part of nourishment construction by preventing them from degradation [30]. It is a promising and rising idea, as the greater part of the biopolymer-based hydrogels should be biodegradable, they can be considered as elective environment friendly packaging materials. Polyvinyl Pyrollidone (PVP) and Carboxymethyl Cellulose (CMC)-based hydrogel was prepared under controlled environmental condition. He prepared different solutions containing PVP and CMC in different concentrations. These concentrations were 20:80, 10:90, 50:50, 80:20, and 90:10. The results showed that CMC-PVP hydrogels of 80:20 ratio was observed useful in packaging of food [32].

12. Various monomers used in IPN hydrogels

See **Table 2.**

Sr no.	Monomer	Monomer abbreviation	Chemical formula	Reference
1	Acrylic acid	AA	$C_3H_4O_2$	[104]
2	Methyl methacrylate	MMA	$CH_2 = C(CH_3)COOCH_3$	[105]
3	Acrylamide	AAm	C_3H_5NO	[106]
4	Itaconic acid	IA	$C_5H_6O_4$	[107]
5	Acrylonitrile	AN	C_3H_3N	[108]
6	Hydroxyethyl methacrylate	HEMA	$C_6H_{10}O_3$	[109]
7	Hydroxypropyl methylcellulose	HPMC	$C_{56}H_{108}O_{30}$	[110]
8	*N*-vinyl 2 pyrolidone	NVP	C_6H_9NO	[111]
9	*N*-isopropyl acrylamide	NIPAAm	$C_6 H_{11}O$	[112]
10	Ethylene glycol	EG	$C_2H_6O_2$	[113]
11	Polyethylene glycol	PEG	$C_{2n}H_{4n+2}O_{n+1}$	[107]

Table 2.
Various monomers used in IPN hydrogels.

13. Patents on hydrogel formulations

See **Table 3**.

Sr no.	Patent	Patent no.	Reference
1	Hydrolysable hydrogels for controlled release	US6497903B1 United States	[114]
2	Hydrolysable hydrogels for controlled release	US7060296B2 United States	[115]
3	Photopolymerizable Biodegradable hydrogels as tissue Contacting materials and Controlled-release carriers	US6,306,922b1	[116]
4	Photopolymerizable biodegradable hydrogels as tissue-contacting materials and controlled-release carriers	US6,306,922b2	[117]
5	Photopolymerizable biodegradable hydrogels as tissue-contacting materials and controlled-release carriers	US6,060,582	[118]
6	Hydrogel compositions for the controlled release administration of growth factors	US6,331,309b1	[119]
7	Catheter with permeable hydrogel membrane	US6,200,257	[120]
8	Controlled-release hydrogel formulation	US20110165236a1 United States	[121]
9	Process for the preparation of a controlled release system	US6303148b1 United States	[122]
10	In situ controlled release drug delivery system	US8940311b2 United States	[123]
11	Lipophilic vehicle-based dual-controlled release matrix system	US8,293,270	[124]
12	Controlled-release hydrogel Formulation	US2008/0075785a1	[125]
13	Polymer-conjugated albumin hydrogels for controlled release of therapeutic agents	US9492376b2 United States	[126]
14	Absorption and controlled release of polyethers from Hydrogel biomaterials	US2004/0115270 a1	[127]
16	Controlled-release oral drug delivery system	US6,692,766	[128]
17	Hydrogel-forming, self-solvating, absorbable polyester copolymers, and methods for use thereof	US 6,413,539	[129]
18	Controlled release hydrogels	US20090104254a1 United States	[130]
19	Hydrophilic vehicle-based dual-controlled release matrix system	Us8333989b2 United States	[131]
20	Controlled release formulations of enzymes, microorganisms, and antibodies with mucoadhesive polymers	Us20080020036a1 United States	[132]
21	Controlled-release formulations	US20130273140A1 United States	[133]
22	Methods of using in situ hydration of hydrogel articles for sealing or augmentation of tissue or vessels	US6605294B2 United States	[134]
23	Controlled-release oral drug Delivery system	US7,189,414b2	[135]

Sr no.	Patent	Patent no.	Reference
24	Chitosan-xanthan-based polyionic hydrogels for stabilization and controlled release of vitamins	US6964772B1 United States	[136]
25	Controlled-release hydrogels	US9999596b2 United States	[137]
26	Controlled release of anti-arrhythmic agents	US20060093673a1 United States	[138]
27	Polymeric devices for controlled Release of active agents	US2006/ 0003008a1	[139]
28	Preparation method of low-pH controlled-release intelligent corrosion inhibitor	US10,131,995b2	[140]
29	Catheter with permeable hydrogel membrane	US6,537,194b1	[141]
30	Polymeric devices for controlled Release of active agents	US8,647,657b2	[142]
31	Self-destructing, controlled-release per-oral drug delivery system	US6365185b1 United States	[143]

Table 3.
Patents on hydrogel formulations.

14. Conclusion

In this book chapter, different types of hydrogel, along with properties as well as medical applications were discussed. However, there are different mechanisms involved in formation of hydrogel. Physically cross-linked nexus employed as substrate for tissue engineering with various medical applications. Drug release from nexus is attributed to swelling. However, stimuli-responsive nexus is gaining much importance. Temperature and pH aid in a controlled release from the hydrogel nexus. Moreover, they have excellent water absorption, soft architecture and are biocompatible.

Author details

Asif Mahmood[1]*, Hira Ijaz[2], Rai Muhammad Sarfraz[3], Nadiah Zafar[4], Muhammad Zaman[5] and Mariya Azam[6]

1 Department of Pharmacy, University of Chakwal, Pakistan

2 Department of Pharmaceutical Sciences, Pak-Austria Fachhochschule: Institute of Applied Sciences and Technology, Haripur, Pakistan

3 College of Pharmacy, University of Sargodha, Sargodha, Pakistan

4 Faculty of Pharmacy, Department of Pharmaceutics, University Teknologi MARA, Malaysia

5 Faculty of Pharmacy, University of Central Punjab, Lahore, Pakistan

6 Department of Pathology, Niazi Medical and Dental College, Sargodha, Pakistan

*Address all correspondence to: asif.mahmood@uoc.edu.pk

References

[1] Mignani S, El Kazzouli S, Bousmina M, Majoral JP. Expand classical drug administration ways by emerging routes using dendrimer drug delivery systems: a concise overview. Advanced Drug Delivery Reviews. 2013;**65**(10):1316-1330

[2] Kaur G, Arora M, Kumar MR. Oral drug delivery technologies—a decade of developments. Journal of Pharmacology and Experimental Therapeutics. 2019; **370**(3):529-543

[3] Campbell S, Smeets N. Drug delivery: polymers in the development of controlled release systems. In: Functional Polymers. Cham: Springer; 2019. pp. 719-747

[4] Hardenia A, Maheshwari N, Hardenia SS, Dwivedi SK, Maheshwari R, Tekade RK. Scientific rationale for designing controlled drug delivery systems. In: Tekade RK, editor. Basic Fundamentals of Drug Delivery. Cambridge, Massachusetts: Academic Press; 2019. pp. 1-28

[5] Park K. Controlled drug delivery systems: past forward and future back. Journal of Controlled Release. 2014;**190**: 3-8

[6] El-Hamshary H et al. Evaluation of clay-ionene nanocomposite carriers for controlled drug delivery: synthesis, in vitro drug release, and kinetics. Materials Chemistry and Physics. 2019; **225**:122-132

[7] Karna S, Chaturvedi S, Agrawal V, Alim M. Formulation approaches for sustained release dosage forms: a review. Asian Journal of Pharmaceutical and Clinical Research. 2015;**8**(5):46-53

[8] Rodríguez-Rodríguez R, Espinosa-Andrews H, Velasquillo-Martínez C, García-Carvajal ZY. Composite hydrogels based on gelatin, chitosan and polyvinyl alcohol to biomedical applications: a review. International Journal of Polymeric Materials and Polymeric Biomaterials. 2020;**69**(1):1-20

[9] Patel GC, Joshi SA. Targeting aspects of hydrogels in drug delivery. In: Grumezescu AM, editor. Biomedical Applications of Nanoparticles. Norwich, NY: William Andrew Publishing; 2019. pp. 195-231

[10] Jagtapa SD, Mohite AM, Tayadeb S, Supriya S. Synthesis of hydrogels and their applications in control fertilizer release. Journal of Emerging Technologies and Innovative Research. 2019;**6**:457-468

[11] Shafiq K, Mahmood A, Salem-Bekhit MM, Sarfraz RM, Algarni AS, Taha EI, et al. Development and Optimization of Tamarind Gum-β-Cyclodextrin-g-Poly (Methacrylate) pH-Responsive Hydrogels for Sustained Delivery of Acyclovir. Pharmaceuticals. 8 Dec 2022; **15**(12):1527

[12] Gupta P, Vermani K, Garg S. Hydrogels: from controlled release to pH-responsive drug delivery. Drug Discovery Today. 2002;**7**(10):569-579

[13] Elsayed MM. Hydrogel preparation technologies: relevance kinetics, thermodynamics and scaling up aspects. Journal of Polymers and the Environment. 2019;**27**(4):871-891

[14] Trotta F, Mele A, editors. Nanosponges: Synthesis and Applications. Hoboken, New Jersey: John Wiley & Sons; 2019

[15] Nayak AK, Hasnain MS. Plant Polysaccharides-Based Multiple-Unit

Systems for Oral Drug Delivery. Singapore: Springer; 2019

[16] Bandawane A, Saudagar R. A review on novel drug delivery system: a recent trend. Journal of Drug Delivery and Therapeutics. 2019;**9**(3):517-521

[17] Chirani N, Yahia LH, Gritsch L, Motta FL, Chirani S, Farè S. History and applications of hydrogels. Journal of Biomedical Science. 2015;**4**(02):1-23

[18] Somya G, Nayyar P, Akanksha B, Kumar SP. Interpenetrating polymer network-based drug delivery systems: emerging applications and recent patents. Egyptian Pharmaceutical Journal. 2015;**14**(2):75

[19] Shivashankar M, Mandal BK. A review on interpenetrating polymer network. International Journal of Pharmacy and Pharmaceutical Sciences. 2012;**4**(5):1-7

[20] Qadri MF, Malviya R, Sharma PK. Biomedical applications of interpenetrating polymer network system. Open Pharmaceutical Sciences Journal. 2015;**2**(1):21-30

[21] Soman A, Mathew F, Chacko AJ, Alias M, Poosan GV. Interpenetrating polymer network (Ipn)-hydrogels. The Pharma Innovations. 2014;**3**(8, Part A): 59-66

[22] Li Y, Wang C, Zhang W, Yin Y, Rao Q. Preparation and characterization of PAM/SA tough hydrogels reinforced by IPN technique based on covalent/ionic crosslinking. Journal of Applied Polymer Science. 2015;**132**(4):1-7

[23] Panahi Y, Gharekhani A, Hamishehkar H, Zakeri-Milani P, Gharekhani H. Stomach-specific drug delivery of clarithromycin using a semi interpenetrating polymeric network hydrogel made of montmorillonite and chitosan: synthesis, characterization and in vitro drug release study. Advanced pharmaceutical bulletin. 2019;**9**(1):159

[24] Wang JJ, Liu F. UV-curing of simultaneous interpenetrating network silicone hydrogels with hydrophilic surface. Polymer Bulletin. 2012;**69**:85-97

[25] Tang Q, Yu JR, Chen L, Zhu J, Hu ZM. Porous silicone hydrogel interpenetrating polymer networks prepared using a template method for biomedical use. Polymer International. 2011;**60**(7):1136-1141

[26] Dragan ES. Design and applications of interpenetrating polymer network hydrogels. A review. Chemical Engineering Journal. 2014;**243**:572-590

[27] Vaghani SS, Patel MM. pH-sensitive hydrogels based on semi-interpenetrating network (semi-IPN) of chitosan and polyvinyl pyrrolidone for clarithromycin release. Drug Development and Industrial Pharmacy. 2011;**37**(10):1160-1169

[28] Bhardwaj V, Harit G, Kumar S. Interpenetrating polymer network (IPN): novel approach in drug delivery. International Journal of Drug Development and Research. 2012;**4**(3): 41-54

[29] Toledo PV, Limeira DP, Siqueira NC, Petri DF. Carboxymethyl cellulose/poly (acrylic acid) interpenetrating polymer network hydrogels as multifunctional adsorbents. Cellulose. 2019;**26**:597-615

[30] Jayaramudu T, Varaprasad K, Sadiku ER, Amalraj J. Temperature-sensitive semi-IPN composite hydrogels for antibacterial applications. Colloids and Surfaces A: Physicochemical and Engineering Aspects. 2019;**572**:307-316

[31] Malikmammadov E, Hasirci N. Dual- and multistimuli-responsive polymers for biomedical applications. In: Aguilar MS, Román JS, editors. Smart Polymers and Their Applications. Sawston, Cambridge: Woodhead Publishing; 2019. pp. 255-278

[32] Lohani A, Singh G, Bhattacharya SS, Verma A. Interpenetrating polymer networks as innovative drug delivery systems. Journal of Drug Delivery. 2014; **2014**:1-11

[33] Amiji MM, inventor; Northeastern University Boston, assignee. Drug delivery using pH-sensitive semi-interpenetrating network hydrogels. United States patent US 5,904,927. 1999

[34] Mukhopadhyay P, Kundu PP. Stimuli-responsive polymers for oral insulin delivery. In: Makhlouf ASH, Abu-Thabit NY, editors. Stimuli Responsive Polymeric Nanocarriers for Drug Delivery Applications. Sawston, Cambridge: Woodhead Publishing; 2019. pp. 525-546

[35] Sano J, Habaue S. Synthesis of interpenetrating polymer networks using silicone polymer with silanol residue. Reactive and Functional Polymers. 2019;**137**:21-26

[36] Hoque J, Sangaj N, Varghese S. Stimuli-responsive supramolecular hydrogels and their applications in regenerative medicine. Macromolecular Bioscience. 2019;**19**(1):1800259

[37] Wang T et al. Biocompatible, injectable anionic hydrogels based on poly (oligo ethylene glycol monoacrylate-co-acrylic acid) for protein delivery. Advanced Therapeutics. 2019;**2**(9):1900092

[38] Fujiyoshi T, Carrez O, Imizcoz M, Zornoza A, Isasi JR. Interpenetrated polymer networks of poly (β-cyclodextrin) and polyvinylpyrrolidone with synergistic and selective sorption capacities. Carbohydrate Polymers. 2019;**219**:105-112

[39] Deen GR, Loh XJ. Stimuli-responsive cationic hydrogels in drug delivery applications. Gels. 2018;**4**(1):13

[40] Olad A, Pourkhiyabi M, Gharekhani H, Doustdar F. Semi-IPN superabsorbent nanocomposite based on sodium alginate and montmorillonite: reaction parameters and swelling characteristics. Carbohydrate Polymers. 2018;**190**: 295-306

[41] Rudhrabatla VP, Jalababu R, Rao KK, Reddy KS. Fabrication and characterisation of curcumin loaded pH dependent sodium alginate-g-poly (acryloyl phenylalanine)-cl-ethylene glycol vinyl ether-co-hydroxyethyl acrylate hydrogels and their in-vitro, in-vivo and toxicological evaluation studies. Journal of Drug Delivery Science and Technology. 2019;**51**:438-453

[42] Hajikhani M, Khanghahi MM, Shahrousvand M, Mohammadi-Rovshandeh J, Babaei A, Khademi SM. Intelligent superabsorbents based on a xanthan gum/poly (acrylic acid) semi-interpenetrating polymer network for application in drug delivery systems. International Journal of Biological Macromolecules. 2019;**139**:509-520

[43] Jalababu R, Veni SS, Reddy KS. Synthesis and characterization of dual responsive sodium alginate-g-acryloyl phenylalanine-poly N-isopropyl acrylamide smart hydrogels for the controlled release of anticancer drug. Journal of Drug Delivery Science and Technology. 2018;**44**:190-204

[44] Kim SJ, Lee CK, Kim IY, Hyeok An K, Kim SI. Preparation and

characterizations of interpenetrating polymer network hydrogels of poly (ethylene oxide) and poly (methyl methacrylate). Journal of Applied Polymer Science. 2003;**89**(1):258-262

[45] Willats WG, Knox JP, Mikkelsen JD. Pectin: new insights into an old polymer are starting to gel. Trends in Food Science & Technology. 2006;**17**(3): 97-104

[46] Cui L, Jia J, Guo Y, Liu Y, Zhu P. Preparation and characterization of IPN hydrogels composed of chitosan and gelatin cross-linked by genipin. Carbohydrate Polymers. 2014;**99**:31-38

[47] Sriamornsak P. Application of pectin in oral drug delivery. Expert Opinion on Drug Delivery. 2011;**8**(8):1009-1023

[48] Fawzy A, Toghan A, Alqarni N, Morad M, Zaki ME, Sanad MM, et al. Experimental and computational exploration of chitin, pectin, and amylopectin polymers as efficient eco-friendly corrosion inhibitors for mild steel in an acidic environment. Kinetic, thermodynamic, and mechanistic aspects. Polymers. 2023;**15**(4):1-33

[49] Singh A, Suri A. Synergistic kinetic hydrate inhibition of pectin, PVP, and PVCap with monoethylene glycol. Energy & Fuels. 2023;**37**(6):4524-4543

[50] Ijaz H, Tulain UR. Development of interpenetrating polymeric network for controlled drug delivery and its evaluation. International Journal of Polymeric Materials and Polymeric Biomaterials. 2019;**68**(18): 1099-1107

[51] Zafar N, Mahmood A, Sarfraz RM, Ijaz H, Ashraf MU, Mehr S. Facile synthesis of β-cyclodextrin-cyclophosphamide complex-loaded hydrogel for controlled release drug delivery. Polymer Bulletin. 2022:1-33. DOI: 10.1007/s00289-022-04567-7

[52] Abdel-Halim ES. Chemical modification of cellulose extracted from sugarcane bagasse: preparation of hydroxyethyl cellulose. Arabian Journal of Chemistry. 2014;**7**(3):362-371

[53] Nayak AK, Pal D. Functionalization of Tamarind Gum for Drug Delivery. In: Thakur V, Thakur M, editors. Functional Biopolymers. Springer, Cham: Springer Series on Polymer and Composite Materials; 2018

[54] Gupta V, Puri R, Gupta S, Jain S, Rao GK. Tamarind kernel gum: an upcoming natural polysaccharide. Systematic Reviews in Pharmacy. 2010;**1**(1):50-54

[55] Dey S, Nandy BC, De JN, Hasnain MS, Nayak AK. Tamarind gum in drug delivery applications. In: Hasnain MS, Nayak AK, editors. Natural Polysaccharides in Drug Delivery and Biomedical Applications. Cambridge, Massachusetts: Academic Press; 2019. pp. 285-306

[56] Malviya R, Sundram S, Fuloria S, Subramaniyan V, Sathasivam KV, Azad AK, et al. Evaluation and characterization of tamarind gum polysaccharide: the biopolymer. Polymers. 2021;**13**(18):3023

[57] Nayak AK, Pal D, Pradhan J, Hasnain MS. Fenugreek seed mucilage-alginate mucoadhesive beads of metformin HCl: design, optimization and evaluation. International Journal of Biological Macromolecules. 2013;**54**:144-154

[58] Gopinath R, Manoharan S, Likhith M, Bhuvan PG, Kailash R, Akash M, et al. Augmentation of soil-moisture retention with natural hydrogel extracted from fenugreek seed husk for loamy soil with real-time monitoring

using FC-28 sensors. ECS Transactions. 2022;**107**(1):1139

[59] Sharma P, Laddha H, Agarwal M, Gupta R. Selective and effective adsorption of malachite green and methylene blue on a non-toxic, biodegradable, and reusable fenugreek galactomannan gum coupled MnO2 mesoporous hydrogel. Microporous and Mesoporous Materials. 2022;**338**:111982

[60] Avachat AM, Dash RR, Shrotriya SN. Recent investigations of plant based natural gums, mucilages and resins in novel drug delivery systems. Indian Journal of Pharmaceutical Education and Research. 2011;**45**(1):86-99

[61] Ijaz H, Tulain UR, Minhas MU, Mahmood A, Sarfraz RM, Erum A, et al. Design and in vitro evaluation of pH-sensitive crosslinked chitosan-grafted acrylic acid copolymer (CS-co-AA) for targeted drug delivery. International Journal of Polymeric Materials and Polymeric Biomaterials. 2022;**71**(5): 336-348

[62] Malatani RT, Bilal S, Mahmood A, Sarfraz RM, Zafar N, Ijaz H, et al. Development of tofacitinib loaded pH-responsive chitosan/mucin based hydrogel microparticles: in-vitro characterization and toxicological screening. Gels. 2023;**9**(3):187

[63] Qureshi J, Iqbal FM, Hussain F, Ijaz H, Aamir MF, Afzal M. Fabrication and evaluation of Olopatadine hydrochloride ocular inserts. Journal of Plastic Film & Sheeting. 2020 Jul;**36**(3):243-259

[64] Mahmood A, Mahmood A, Sarfraz RM, Ijaz H, Zafar N, Ashraf MU. Hydrogel-based intelligent delivery system for controlled release of diloxanide furoate. Polymer Bulletin. 2022;**1**:1-37

[65] Batool N, Mahmood A, Sarfraz RM, Ijaz H, Zafar N, Hussain Z. Formulation and evaluation of interpenetrating polymeric network for controlled drug delivery. Drug Development and Industrial Pharmacy. 2021;**47**(6): 931-946

[66] López-Marcial GR, Zeng AY, Osuna C, Dennis J, García JM, O'Connell GD. Agarose-based hydrogels as suitable bioprinting materials for tissue engineering. ACS Biomaterials Science & Engineering. 2018;**4**(10):3610-3616

[67] Salati MA, Khazai J, Tahmuri AM, Samadi A, Taghizadeh A, Taghizadeh M, et al. Agarose-based biomaterials: opportunities and challenges in cartilage tissue engineering. Polymers. 2020;**12** (5):1150

[68] Hur J, Im K, Kim SW, Kim J, Chung DY, Kim TH, et al. Polypyrrole/agarose-based electronically conductive and reversibly restorable hydrogel. ACS Nano. 2014;**8**(10):10066-10076

[69] Martínez-Sanz M, Ström A, Lopez-Sanchez P, Knutsen SH, Ballance S, Zobel HK, et al. Advanced structural characterisation of agar-based hydrogels: Rheological and small angle scattering studies. Carbohydrate Polymers. 2020; **236**:115655

[70] Sarfraz RM, Ahmad M, Mahmood A, Ijaz H. Development, in vitro and in vivo evaluation of pH responsive β-CD-comethacrylic acid-crosslinked polymeric microparticulate system for solubility enhancement of rosuvastatin calcium. Polymer-Plastics Technology and Engineering. 2018;**57**(12):1175-1187

[71] Huang Z, Wu Q, Liu S, Liu T, Zhang B. A novel biodegradable β-cyclodextrin-based hydrogel for the removal of heavy metal ions. Carbohydrate Polymers. 2013;**97**(2):496-501

[72] Daoud-Mahammed S, Grossiord JL, Bergua T, Amiel C, Couvreur P, Gref R. Self-assembling cyclodextrin based hydrogels for the sustained delivery of hydrophobic drugs. Journal of Biomedical Materials Research Part A. 2008;**86**(3):736-748

[73] Tan H, DeFail AJ, Rubin JP, Chu CR, Marra KG. Novel multiarm PEG-based hydrogels for tissue engineering. Journal of Biomedical Materials Research Part A: An Official Journal of The Society for Biomaterials, The Japanese Society for Biomaterials, and The Australian Society for Biomaterials and the Korean Society for Biomaterials. 2010;**92**(3):979-987

[74] Weber LM, He J, Bradley B, Haskins K, Anseth KS. PEG-based hydrogels as an in vitro encapsulation platform for testing controlled β-cell microenvironments. Acta Biomaterialia. 2006;**2**(1):1-8

[75] Sagle AC, Ju H, Freeman BD, Sharma MM. PEG-based hydrogel membrane coatings. Polymer. 2009;**50**(3):756-766

[76] Shaw GS, Uvanesh K, Gautham SN, Singh V, Pramanik K, Banerjee I, et al. Development and characterization of gelatin-tamarind gum/carboxymethyl tamarind gum based phase-separated hydrogels: A comparative study. Designed Monomers and Polymers. 2015;**18**(5):434-450

[77] Pushpamalar J, Meganathan P, Tan HL, Dahlan NA, Ooi LT, Neerooa BN, et al. Development of a polysaccharide-based hydrogel drug delivery system (DDS): an update. Gels. 2021;7(4):153

[78] Sennakesavan G, Mostakhdemin M, Dkhar LK, Seyfoddin A, Fatihhi SJ. Acrylic acid/acrylamide based hydrogels and its properties - a review. Polymer Degradation and Stability. 2020;**180**: 109308

[79] Tang Y, Hu M, Tang F, Huang R, Wang H, Wu D, et al. Easily-injectable shear-thinning hydrogel provides long-lasting submucosal barrier for gastrointestinal endoscopic surgery. Bioactive Materials. 2022;**15**:44-52

[80] Ijaz H, Tulain UR, Qureshi J. Formulation and in vitro evaluation of pH-sensitive cross-linked xanthan gum-grafted acrylic acid copolymer for controlled delivery of perindopril erbumine (PE). Polymer-Plastics Technology and Engineering. 2018;**57** (5):459-470

[81] Gao Z, Golland B, Tronci G, Thornton PD. A redox-responsive hyaluronic acid-based hydrogel for chronic wound management. Journal of Materials Chemistry B. 2019;7(47): 7494-7501

[82] Park YD, Tirelli N, Hubbell JA. Photopolymerized hyaluronic acid-based hydrogels and interpenetrating networks. Biomaterials. 2003;**24**(6): 893-900

[83] Jaipan P, Nguyen A, Narayan RJ. Gelatin-based hydrogels for biomedical applications. MRS Communications. 2017;7(3):416-426

[84] Kang JI, Park KM. Advances in gelatin-based hydrogels for wound management. Journal of Materials Chemistry B. 2021;**9**(6):1503-1520

[85] Zheng Y, Andreopoulos FM, Micic M, Huo Q, Pham SM, Leblanc RM. A novel photoscissile poly (ethylene glycol)-Based Hydrogel. Advanced Functional Materials. 2001;**11**(1):37-40

[86] Park Y, Lutolf MP, Hubbell JA, Hunziker EB, Wong M. Bovine primary chondrocyte culture in synthetic matrix metalloproteinase-sensitive poly (ethylene glycol)-based hydrogels as a

scaffold for cartilage repair. Tissue Engineering. 2004;**10**(3–4):515-522

[87] Kumar A, Han SS. PVA-based hydrogels for tissue engineering: a review. International Journal of Polymeric Materials and Polymeric Biomaterials. 2017;**66**(4):159-182

[88] Husain MS, Gupta A, Alashwal BY, Sharma S. Synthesis of PVA/PVP based hydrogel for biomedical applications: a review. Energy Sources, Part A: Recovery, Utilization, and Environmental Effects. 2018;**40**(20): 2388-2393

[89] Roy N, Saha N, Kitano T, Saha P. Biodegradation of PVP–CMC hydrogel film: A useful food packaging material. Carbohydrate Polymers. 2012;**89**(2):346-353

[90] Doan P, Nhi TT, Nguyen DT, Nguyen BT, Nguyen TP, Tran NQ. Multifunctional injectable pluronic-cystamine-alginate-based hydrogel as a novel cellular delivery system towards tissue regeneration. International Journal of Biological Macromolecules. 2021;**185**: 592-603

[91] Xiong XY, Tam KC, Gan LH. Polymeric nanostructures for drug delivery applications based on Pluronic copolymer systems. Journal of Nanoscience and Nanotechnology. 2006; **6**(9–10):2638-2650

[92] Engel L, Berkh O, Adesanya K, Shklovsky J, Vanderleyden E, Dubruel P, et al. Actuation of a novel pluronic-based hydrogel: Electromechanical response and the role of applied current. Sensors and Actuators B: Chemical. 2014;**191**: 650-658

[93] Ijaz H, Tulain UR, Azam F, Qureshi J. Thiolation of arabinoxylan and its application in the fabrication of pH-sensitive thiolated arabinoxylan grafted acrylic acid copolymer. Drug Development and Industrial Pharmacy. 2019;**45**(5):754-766

[94] Murphy N, Lynch MA. Activation of the P 2 X 7 receptor induces migration of glial cells by inducing cathepsin B degradation of tissue inhibitor of metalloproteinase 1. Journal of Neurochemistry. 2012;**123**(5):761-770

[95] Peppas NA, Bures P, Leobandung WS, Ichikawa H. Hydrogels in pharmaceutical formulations. European Journal of Pharmaceutics and Biopharmaceutics. 2000;**50**(1):27-46

[96] Lin F, Lu X, Wang Z, Lu Q, Lin G, Huang B, et al. In situ polymerization approach to cellulose–polyacrylamide interpenetrating network hydrogel with high strength and pH-responsive properties. Cellulose. 2019;**26**:1825-1839

[97] Domb AJ, Kost J, Wiseman D, editors. Handbook of Biodegradable Polymers. CRC Press; 1998

[98] Bera R, Dey A, Chakrabarty D. Synthesis and characterization of acrylic acid-2-hydroxyethyl methacrylate IPN hydrogels. RSC Advances. 2015;5(93): 75870-75880

[99] Kampa GJ. Patent No. WO 2006/ 138461 A3. Alexandria Virgenia, Patent Cooperation Treaty. 2006

[100] Murphy SV, Skardal A, Atala A. Evaluation of hydrogels for bio-printing applications. Journal of Biomedical Materials Research Part A. Jan 2013;**101** (1):272-284

[101] Zhang N, Shen Y, Li X, Cai S, Liu M. Synthesis and characterization of thermo-and pH-sensitive poly (vinyl alcohol)/poly (N, N-diethylacrylamide-

co-itaconic acid) semi-IPN hydrogels. Biomedical Materials. 2012;7(3):035014

[102] Mohamed RR, Seoudi RS, Sabaa MW. Synthesis and characterization of antibacterial semi-interpenetrating carboxymethyl chitosan/poly (acrylonitrile) hydrogels. Cellulose. 2012;**19**:947-958

[103] Sun XF, Zeng Q, Wang H, Hao Y. Preparation and swelling behavior of pH/temperature responsive semi-IPN hydrogel based on carboxymethyl xylan and poly (N-isopropyl acrylamide). Cellulose. 2019;**26**:1909-1922

[104] Lee DS, Kang JI, Hwang BH, Park KM. Interpenetrating polymer network hydrogels of gelatin and poly (ethylene glycol) as an engineered 3D tumor microenvironment. Macromolecular Research. 2019;**27**:205-211

[105] Das S, Subuddhi U. Controlled delivery of ibuprofen from poly (vinyl alcohol)– poly (ethylene glycol) interpenetrating polymeric network hydrogels. Journal of Pharmaceutical Analysis. 2019;**9**(2):108-116

[106] Hennink WE, Franssen O. U.S. Patent No. 6,303,148. Washington, DC: U.S. Patent and Trademark Office. 2001

[107] Hennink WE, van Dijk-Wolthuis WNE. U.S. Patent No. 6,497,903. Washington, DC: U.S. Patent and Trademark Office. 2002

[108] Hubbell JA, Kieras MT, Ron ES, Rowe SC. U.S. Patent No. 6,703,037. Washington, DC: U.S. Patent and Trademark Office. 2004

[109] Hubbell JA, Pathak CP, Sawhney AS, Desai NP, & Hill-West JL. U.S. Patent No. 6,060,582. Washington, DC: U.S. Patent and Trademark Office. 2000

[110] Hennink WE, van Dijk-Wolthuis WNE. U.S. Patent No. 7,060,296. 2006

[111] Jennings Jr RN, Yang B, Protter AA, Wang YCJ. U.S. Patent No. 6,331,309. Washington, DC: U.S. Patent and Trademark Office. 2001

[112] Winkler RA. U.S. Patent No. 6,200,257. Washington, DC: U.S. Patent and Trademark Office. 2001

[113] Chow SL, Wong D. U.S. Patent Application No. 13/032,453. 2011

[114] Lim TH, Park JB, Lee JW. U.S. Patent No. 8,940,311. Washington, DC: U.S. Patent and Trademark Office. 2015

[115] Sukuru K. U.S. Patent No. 8,333,989. Washington, DC: U.S. Patent and Trademark Office. 2012

[116] Chow SL, Wong D, Garcia D. U.S. Patent Application No. 11/772,017. 2008

[117] Seliktar D, Oss-Ronen L. U.S. Patent No. 9,492,376. Washington, DC: U.S. Patent and Trademark Office. 2016

[118] Jani D, Salamone J, Hu Z, Xia E, Borazjani R, Ammon D. U.S. Patent Application No. 10/319,132. 2004

[119] Rubinstein A, Radai R, Friedman M, Tirosh B, Baluom M, & Nassar T. U.S. Patent No. 6,692,766. Washington, DC: U.S. Patent and Trademark Office. 2004

[120] Shalaby SW. U.S. Patent No. 6,413,539. Washington, DC: U.S. Patent and Trademark Office. 2002

[121] Sinko PJ, Stein S, Lalloo A.. U.S. Patent Application No. 11/793,566. 2009

[122] Sukuru K. U.S. Patent No. 8,293,270. Washington, DC: U.S. Patent and Trademark Office. 2012

[123] Jolly J. U.S. Patent Application No. 11/812,638. 2008

[124] Maggio ET, Meezan E, Pillion DJ, Morgan SL, Baggott J. U.S. Patent Application No. 13/915,604. 2013

[125] Sawhney AS, Ransone IWH. U.S. Patent No. 9,775,906. Washington, DC: U.S. Patent and Trademark Office. 2017

[126] Rubinstein A, Friedman M, Baluom M, Tirosh B. U.S. Patent No. 7,189,414. Washington, DC: U.S. Patent and Trademark Office. 2007

[127] Chornet E, Dumitriu S. U.S. Patent No. 6,964,772. Washington, DC: U.S. Patent and Trademark Office. 2005

[128] Sinko PJ, Stein S, Lalloo A. U.S. Patent No. 9,999,596. Washington, DC: U.S. Patent and Trademark Office. 2018

[129] Coury A, Philbrook C, Skinner K. U.S. Patent Application No. 10/533,179. 2006

[130] Gibson J, Tipton A. U.S. Patent Application No. 11/022,416. 2006

[131] Li X, Dong C, Cheng X, Du C, Xiao K. U.S. Patent No. 10,131,995. Washington, DC: U.S. Patent and Trademark Office. 2018

[132] Winkler RA. U.S. Patent No. 6,537,194. Washington, DC: U.S. Patent and Trademark Office. 2003

[133] Gibson JW, Tipton AJ. U.S. Patent No. 8,647,657. Washington, DC: U.S. Patent and Trademark Office. 2014

[134] Ritschel WA, Agrawal MA. U.S. Patent No. 6,365,185. Washington, DC: U.S. Patent and Trademark Office. 2002

[135] Kuzma P, Decker S. U.S. Patent No. 7,452,868. Washington, DC: U.S. Patent and Trademark Office. 2008

[136] Hussain MA, Repta AJ. U.S. Patent No. 6,649,187. Washington, DC: U.S. Patent and Trademark Office. 2003

[137] Maggio ET, Meezan E, Pillion DJ, Morgan SL, Baggott J. U.S. Patent No. 8,329,220. Washington, DC: U.S. Patent and Trademark Office. 2012

[138] Hubbell JA, Pathak CP, Sawhney AS, Desai NP, Hill JL. U.S. Patent No. 6,602,975. Washington, DC: U.S. Patent and Trademark Office. 2003

[139] Sawhney AS. U.S. Patent No. 6,605,294. Washington, DC: U.S. Patent and Trademark Office. 2003

[140] Dave BC. U.S. Patent Application No. 12/180,561. 2009

[141] Won CY, Zhang Y, Chu CC. U.S. Patent No. 6,916,857. Washington, DC: U.S. Patent and Trademark Office. 2005

[142] Tabata Y. U.S. Patent No. 8,771,734. Washington, DC: U.S. Patent and Trademark Office. 2014

[143] Chen CM, Xie J, Jan S.. U.S. Patent No. 6,210,716. Washington, DC: U.S. Patent and Trademark Office. 2001

Chapter 4

Physiochemical and Biomedical Properties of Hydrogels: From Fundamentals to Applications

Ruby Varghese, Yogesh Bharat Dalvi, P. Lochana, S. Achinthya, Bhagyashri Omprakash Somani, Preetha Karnaver, Nebu George Thomas, S. Rupesh, Nibu Varghese and Jayachandran V.P.

Abstract

Translational research is utilizing the hydrophilic characteristic of polymer structures, which possess the physical or chemical cross-linking capability. This attribute has been applied in pharmaceutical research to develop hydrogels, which are increasingly being utilized for cell and drug delivery, soft and hard tissue regeneration, wound healing, regenerative medicine, contrast imaging, radiation shielding, and enhancing the biocompatibility of clinical implants. This chapter concentrates on the physicochemical and mechanical characteristics of hydrogels, such as surface properties, contact angle, tensile strength, and swelling behavior, and how these properties affect the biodegradability, stimuli sensitivity, and biomedical uses of hydrogels. Ultimately, this review provides readers with an overview of the advancements and challenges in each segment, albeit not all pertinent issues can be explored in detail due to the intricacy of biological responses to the hydrogel.

Keywords: biocompatibility, drug delivery, hydrogel-based scaffolds, regenerative medicine, surface properties, tissue engineering

1. Introduction

Biomaterials referred to as "hydrogels" are a class of polymeric materials that can hold much water in their dynamic network due to their hydrophilic structure. Though it holds a considerable amount of water in its structure, the crosslinked network does not dissolve in it [1]. Ideally, hydrophilic units are polymerized to produce hydrogel, but occasionally hydrophobic units are also used to control the properties associated with a particular application of interest. The water absorption attribute gives hydrogels a considerable degree of elasticity akin to actual tissue in a variety of biomedical applications [2]. Since its emergence as a biomedical tool six decades ago, its application has been increasing tremendously due to its surface behavior, biocompatibility, diffusive properties, biodegradability, tensile strength, and responsiveness to cues.

Due to these properties, hydrophilic polymeric networks are extensively used in biomedical engineering of tissue [3], drug and gene loading [4], cell delivery [5], and prostheses [6] and also in dermatology. A recent advancement in hydrogels, smart hydrogels have the ability to react to environmental signals. This enables the tailored use of hydrogels in several domains. These gels react to various physical and chemical cues finding applications in regenerative treatment, biological isolation, neural stimulation, biosensing, and many more.

This review tries to focus on various biomedical applications of hydrogel due to the fundamental properties they possess.

2. Structure of hydrogels

Hydrogels are polymers formed by repeated hydrophilic units that can take in large amounts of water. Hydrogel's polymer chains are connected by cross-linking to form a virtually limitless network, preventing the chains from dissolving in a fluid medium. The interaction between the hydrophilic unit and water can take place through polar interactions, ionic interactions, or hydrogen bonding [5]. The structural arrangement of the hydrogel is crucial in the analysis of associated characteristics.

Based on the structure and conformation of the starting material used, the connecting macromolecular chain forms the sol [7], which further leads to gelation. The branching polymer size increases as the cross-linking process continues. The branching polymer's size is inversely proportional to its solubility. Some of the hydrophilic groups present in the hydrogel network that accounts for its water absorption capacity are COOH, OH, NH2, CONH, CONH2, and SO3H [8]. Depending on the specific polymers employed and the crosslinking density, hydrogel structures can change. Hydrogels are generally divided into two groups: natural hydrogels and synthetic hydrogels [9–11].

Naturally occurring polymers, such as polysaccharides and proteins, form the basis of natural hydrogels. These hydrogels frequently replicate the extracellular matrix (ECM) present in real tissues and create an environment that encourages cell attachment and development. Natural hydrogels often have an uneven network structure with varying pore size and shape. The porous structure facilitates the movement of nutrients and gaseous exchange. Naturally derived hydrogels are advantageous owing to their intrinsic biodegradability and biocompatibility. This makes them ideal for biomedical purposes [9, 10]. Synthetic hydrogels are engineered polymers to produce desired characteristics unlike naturally occurring polymers [12]. Although they lack inherent bioactivity, they have a wide range of applications [9, 10].

3. Fundamental characteristics of hydrogels

Surface properties: The biocompatibility of a substance is greatly influenced by its surface chemistry. The foremost point of contact between the hydrogel and the surrounding tissue system is its surface. The physicochemical and topographical surface features of the hydrogel are crucial factors in regulating and influencing cellular adhesion and proliferation. In order to comprehend how surface chemistry affects tissue response, surface qualities, such as hydrophilic traits, surface charge, and surface functionality, have been intensively studied [13]. In order to accommodate the number of cells needed to replace or restore tissue functions, a hydrogel with relatively broad and accessible surface areas is employed. Hydrogel scaffold surface properties can be selectively enhanced using a variety of techniques. These surface changes could lead to enhanced biocompatibility and specificity [14]. The surface of hydrogels can range from crystalline to disordered matrices, or from rough to smooth. Similar to the extracellular matrix in natural tissues, these organizational and structural characteristics have been observed to have a substantial impact on the fate of the cells during the tissue regeneration process [15].

Swelling behavior: Due to the thermodynamic compatibility of the polymer chains with hydrophilic units and water, the biopolymer network begins to swell when it comes into contact with an aqueous solution or a biological fluid [9, 10]. The network's cross-links cause a reverse force that balances off the swelling force. When these two forces are balanced, swelling equilibrium is attained. The swelling behavior of hydrogel is crucial for determining factors [12], including the degree of cross-linking, mechanical characteristics, and rate of disintegration [16]. The swelling equilibrium varies depending on the monomer hydrophilicity. The higher the hydrophilicity of the monomer involved in forming the polymer, the higher the water absorption (**Table 1**) [14].

Biocompatibility: Every biomaterial that interacts with tissue needs to be biocompatible to achieve its therapeutic effectiveness. For any application, it is essential to comprehend how the hydrogel behaves in a biological system, in particular the characteristics of its interactions with adjacent tissue [20]. Biocompatibility is broadly classified into bulk and interfacial biocompatibility. The ability of a material to impose physiological and mechanical stimulation on the systems that it surrounds

Natural polymers	**Source**	**Properties**	**Limitations**
Alginate	Derived from plants, animals, microorganisms, and algae.	Biocompatibility, biodegradability, biosafety, low immunogenicity, cost-effective, and adhesive in nature.	Mechanical properties are limited, difficulty of purification, and minor immune response due to impurities.
Agarose			
Chitosan			
Collagen			
Cellulose			
Fibrin			
Gelatin			
Hyaluronic			
pectin			
Synthetic polymers	Source	Properties	Limitation
PEG [Polyethylene]	Polymerization of synthetic monomers	Economical than natural polymers, prolonged shelf life, efficiently delivers soluble molecules, unreactive, and degradation rate can be regulated and highly reproducible.	Triggers immune response, triggers inflammation reactions, and low biocompatibility.
PVA [poly (vinyl alcohol)]			
PU [Polycarbonate urethane]			
Poly [epsilon-caprolactone]			
Poly [anhydride]			
PPF [Propylene fumarate]			
PCL [Poly (caprolactone)]			
PLA [Poly (lactic acid)]			
PLGA [Poly (lactic-co-glycolic acid)]			

Table 1.
Various natural and synthetic polymers used in hydrogels, their source, properties, and limitations [17–19].

is referred to as bulk biocompatibility, also known as mechanical biocompatibility. Interactions between the material and its biological environment are accounted for by the processes of protein adsorption and cell adhesion in interfacial biocompatibility. Biocompatibility for biomedical purposes appears to have more to do with interfacial compatibility than bulk compatibility [21, 22].

Diffusive properties: Numerous uses of hydrogel in bioengineering are based on the capacity to control solute transport through them. The process of diffusion is significantly different in polymers of hydrogel when compared to small molecules. The diffusion process is majorly impacted by the interactions between solutes, gel polymers, and solvents. Diffusion and macromolecular relaxation work together to govern drug release in swelling-controlled systems, resulting in zero-order release circumstances [23]. Diffusion-mediated hydrogel systems can be reservoir or matrix systems. A hydrogel membrane allows the active substance to diffuse before reaching the biological fluid. The reservoir system's active agent is situated in a core and is encircled by a polymer membrane. In matrix systems, the medication or protein is uniformly dispersed over the membrane and liberated over time [24, 25].

Biodegradability: The hydrogel structure contains labile connections that lead to their breakdown in the aqueous condition or on the action of enzymes and are regulated by various external and internal factors, finally causing their degradation. The degree and rate of biodegradation of hydrogels are crucial in tissue engineering. Since hydrogels serve as a medium for the growth of the tissues, they eventually must undergo degradation [26]. Cells need room to proliferate, therefore, hydrogel degradation must perfectly coincide with cell multiplication during tissue regeneration. The time of hydrogel degradation determines the success of tissue engineering [15]. For drug release studies, constant monitoring of degradation is required. An early degradation might be triggered under nutrient-deficient conditions or degradation could be delayed further leading to immune responses. According to recent studies, by changing the gel composition or with the use of a laser, controlled degradation of the hydrogel may be achieved [27].

Stimuli sensitivity: Stimuli-sensitive hydrogels are hydrogels that physically and chemically respond to certain environmental factors. Swelling behavior of the hydrogel changes in response to the stimuli [28]. The stimuli could be endogenous or exogenous [29]. Endogenous stimuli are the ones present in the bio-environment of the hydrogel produced by the organism, while exogenous stimuli are external cues. Endogenous factors include metal ion availability, enzymes, pH, antigen, etc., and exogenous factors include temperature, light, magnetic field, electric field, and others. Tissue engineering studies are increasing with the evolution of stimuli-responsive hydrogels. Exogenous stimuli-responsive hydrogels are majorly employed for various biomedical applications to obtain desired results [29]. These stimuli-sensitive hydrogels are known as "Smart hydrogels" [15], which will be dealt in detail in the later part of the review.

4. Applications

4.1 Hydrogels in tissue engineering

The field of tissue engineering and regenerative medicine has tremendous scope for advanced treatment outcomes with pioneering bioengineering technologies. Hydrogels have played a pivotal role, as a tissue-engineered scaffold over the past few decades due to their multifarious physiochemical properties, such as mechanical rigidity, biodegradability, swellability, biocompatibility, and stimuli sensitivity. It also provides an ideal niche for cell survival, cell proliferation, differentiation, and migration, thereby mimicking the native tissue.

4.2 Exigency and significance of hydrogel-based scaffold

A. Human cell types require an ideal anchorage to support the regeneration of tissues, lacking, which may result in malfunctioned tissues and cell necrosis.

B. An ideal scaffold should not only act as an anchorage but should also provide the native environment and exhibit limited interaction with stromal cells which is crucial for tissue regeneration [30].

Hydrogels are 3D natural or synthetic hydrophilic polymer network, which has been designed to overcome the limitations of conventional scaffolds used in biomedical applications. Due to higher water content, it does not get dissolved in a high concentration of water, mimics the native environment, helps in the diffusion of nutrients, and provides biochemical and structured support to the surrounding cells catering to a platform for the tissue to function properly without affecting its overall appearance. Hence, hydrogel is considered as an innovative and novel material for tissue engineering and regenerative purposes. Numerous amounts of hydrogels are constructed to be utilized for clinical purposes, such as stimuli-responsive hydrogels, which include physical-responsive hydrogels (temperature, electro, and magnetic-responsive hydrogels); and chemical-responsive hydrogels (pH, glucose, and biological/biochemical-responsive hydrogel). Smart hydrogels are employed in biomedical and health sectors due to their unique ability to modulate physiochemical and mechanical properties to fit the desired application. Hydrogels are commonly employed in areas, such as bone, cartilage, meniscus, vascular tissue, tendon, eyes, and soft tissue. Hydrogel-mediated gene, cell, and drug delivery play a prominent role in regenerative medicine and tissue engineering [31].

4.3 Hydrogel in gene delivery

The hydrogel scaffolds with their properties of swellability and mild gelation provide an appropriate condition for the transport of nucleic acid without any hindrance to the targeted tissues. Various authors have reported their design, which can transport any genes (viral or non-viral) to the destined site by avoiding clearing mechanisms, controlled release *via* provision of cellular migration, and differentiation. Komatsu et al. developed a gelatin-collagen-based plasmid DNA delivery for the induction of bone regeneration, which seems to be more efficient than atelocollagen as substrate [32]. Similarly, in various other tissues, such as cartilage regeneration, tendon injury, cardiovascular tissue repair, skin tissue repair, and nervous tissue repair hydrogels, are employed. Due to their biocompatibility and biosafety, these materials are used in delivering angiogenic factors [33] for cardiovascular diseases and RNAi drug product on patrol and SiRNA for the hereditary transthyretin amyloidosis approved in 2018 [34, 35].

Delivery of genetic material by smart biomaterial provides a unique chance to capitalize the synergistic interaction between the hydrogel and genetic vectors for cellular process and gene delivery. Smart hydrogels can be tuned in such a manner that it can directly control the events from cell engraftment to its delivery at targeted site that includes the extent of infiltration of cell and preservation of vector activity and its retention [36].

4.4 Hydrogels in drug delivery

The ability to tailor the properties of hydrogel-based scaffolds during production and their applicability for safe implantation, release, and degradation makes them appealing for controlled drug delivery [8]. It's a crosslinked polymer that can be used for delivering drug *via* various routes of administration which include oral, nasal, rectal, ocular, parenteral, topical, orthotopic, intraperitoneal, and transdermal [4]. Some of the routes utilized by hydrogels for drug delivery are mentioned below:

1. **Subcutaneous:** Various hydrogels have been administered to the immune-privileged subcutaneous tissue to evaluate the therapeutic response and to assess its toxicity, such as polyethylene hydrogels, ellagic acid hydrogels, nano-patterned polyacrylamide hydrogels, chitosan, alginate, pectin, and gelatin hydrogels, most of it has displayed mild to negligible inflammatory responses.

2. **Oral:** Oral-administered hydrogels, such as MPEG, caprolactone, and itaconic acid, are pH-sensitive hydrogel; and photo-polymerized pH-responsive hydrogels have shown to be nontoxic, but their effectiveness is limited because of:

   A. Cleavage by digestive enzymes.

   B. Less diffusability from epithelial membrane to bloodstream.

3. **Rectal:** Rectal delivery has an advantage over oral delivery due to rapid absorption rate, avoidance of digestive enzymes, and provision of control release with limited or no adverse relations. Many hydrogels such as catechol-chitosan-based hydrogel, mucoadhesive chitosan hydrogel; hydrogel spacer hydrogel are suppositories for rectal tumors.

4. **Ocular:** Various stimuli-responsive hydrogels are used for ocular delivery, such as temperature-sensitive, pH-sensitive, ion-sensitive, ultrasound responsive, and hydrogel-based iontophoresis, to overcome the challenge observed in conventional ophthalmic treatments.

4.5 Nanoparticle-loaded hydrogel in drug delivery

The administration of drugs *via* nanoparticles has been employed for quite a while. In spite of its advantages like durability, biodegradability, and ability to transport both hydrophobic and hydrophilic drug [37, 38], there are certain limitations, including premature release of drug, instability of nanoparticles upon contact with bodily environment, and immune system clearance. These limitations are effectively reduced by incorporating drug-loaded nanoparticles into the hydrogel matrix. Hydrogels readily encapsulate molecules, and shield and aid in releasing them gradually, while also raising their localization and lowering harmful effects in neighboring tissues. Due to its behavior during the sol-gel transition at body temperature, thermo-sensitive hydrogels are the generally employed type of hydrogel for medical applications [39]. The fluid suspension injected [40] into the body quickly transforms into a stable gel network at body temperature [41]. It is possible for these hydrogels to perfectly conform to the geometry of the area where they are applied, resulting in the creation of a drug repository for a controlled and sustained release. Combining NPs and hydrogels into a single system would enable their individual limitations to be covered [42].

4.6 Hydrogels in cell delivery

Hydrogels are extensively used as a scaffold to deliver cells and for various biomedical applications. This matrix provides a platform for the modest engraftment of transplanted cells to the target site by providing cell protection, enhancement, and prolonged retention as these hydrophilic polymers mimic native ECM [5].

Two important strategies *via* which hydrogels deliver cells are:

1. Encapsulation of the cells from the host tissue by the hydrogel through non-integrating approaches.

2. Implementing an integrative approach to have direct contact between transplanted cells and host tissue through biodegradation or microporous design.

4.7 Cell encapsulated hydrogels

Cells can be encapsulated into hydrogel *via* various techniques, such as emulsion, electrostatic droplet extrusion, lithography, lithography, 3D printing, and miro molding. Cell encapsulation facilitates cells in an environment to carry out normal functions by providing an immune regulatory barrier for the better survival of transported cells.

Mesenchymal stem cells and its derivative secretome have potential therapeutic relevance, but meticulous research is lacking in these types of cells due to low retention and poor survival rate. Hence, the hydrogel-encapsulated MSCs boost the treatment of MSCs and their derivatives to the next level due to their improved directional delivery and promotion of their therapeutic behavior.

Gao et al. [43] constructed MSC-loaded bioglass/γ- poly glutamic acid/chitosan hydrogel, which evoked active interaction between MSCs and cellular matrix inducing angiogenesis, improving cardiac function, mitigating cardiac remodeling, and decreased apoptosis.

Hydrogels are multifaceted polymers due to their unique property, which makes them an exemplary candidate for wound healing applications [43, 44], demonstrated that thermostable injectable chitosan/collagen/β-glycerophosphate hydrogel-encapsulated MSCs enhance chronic wound healing caused by venous diseases or diabetes. Hydrogel displays this potential therapeutic activity due to its capability to maintain cell morphology and viability, noninterference with the bioactivity of deliverable cells with no adverse reactions [44].

4.8 Hydrogel in prosthetic and orthodontic implants

Hydrogel is a versatile material in dentistry. In periodontology, utilization of biocompatible membranes is of utmost necessity as implantation of this membrane around diseased periodontium, will prevent unwanted migration of soft tissue and provide a suitable environment and time for regeneration of bones, tissues, and ligaments. Commercially available scaffolds are having limitations, such as being non-resorbable and non-biocompatible. Hence, the utilization of hydrogels has proven to be effective as they are biocompatible, form a 3D network *in situ* once injected, and prevent invasion by unwanted tissues [45]. Hydrogels have been extensively researched in the area of prosthodontics. Since they are bioinert, non-toxic, and stable in light, they are used for the maxillofacial area, which is continuously exposed to sunlight [46]. Due to its flexible nature, physical and chemical modifications carried out in the hydrogel make it a definitive tissue-engineered scaffold to be used as maxillofacial prosthetic material.

Further, due to its variable ability to carry cells into the root canal system, it mimics ECM and provides the niche for the new cells to grow. Hence, hydrogel scaffold is preferred in endodontics [47, 48].

4.9 Hydrogel in dermatology

Hydrogels are fascinating group of designed polymeric materials to be used in dermatology.

In dermatology, hydrogels are preferred for transdermal applications, as they are easy to apply, cause negligible to minimal adverse effects, have no sudden elevation in serum concentrations, have resemblance to the bodily tissues, possess numerous sites for modification, less immunogenicity, and enhanced effectiveness. Hydrogels have continuously been used in treating various anomalies in the form of self-adhesive patch, microcapsule-embedded hydrogel patches, injectable shape memoizable 3D Hyaluronic acid cryogels, peel-off hydrogel masks, etc., as these are effective and support skin regeneration due to their ability to promote of drug penetration, mitigate fungal growth, reduction of lesions, and enhanced dry permeability with visible clinical improvement [49].

4.10 Smart hydrogels: Concept and applications

Recently, significant progress has been made in the development and study of a special class of hydrogels known as smart hydrogels [50]. In response to diverse environmental stimuli, including pH, temperature, light, ionic strength, magnetic field, and electric field, smart hydrogel displays remarkable variations in their swelling behavior, network configuration, and physical characteristics [37, 38, 51]. The changes in the structure of the hydrogel are in accordance with the magnitude of the signal received. When smart hydrogels are exposed to any of these stimuli, they undergo changes that are typically reversible. When the stimulus is removed, the hydrogels revert to their earlier state [52].

Due to the stimuli-responsive property of these hydrogels, they have potential role in various biomedical applications [53]. Most of the techniques are still in developmental stages, some of the established applications are discussed below.

Hydrogel-based wearable biosensor: A biosensor is a device that detects the signal of interest in its physiological environment [54]. The biomedical sciences are currently developing many applications for wearable biosensors. These wearable biosensors are conjugated with hydrogels, due to their similarity with tissue softness, hydration, compatibility, and distinctive ionic sensing capabilities. The differences between human bodies and the conventional electrical akin are significantly reduced by the use of hydrogels. For wearable sensing, a number of hydrogel-derived, skin-like devices are now available. These devices can replicate the sensory capabilities of the skin by converting inputs, such as touch, pressure, humidity, and temperature, into detectable variations in electrical impulses [55].

Numerous wearable epidermal biosensors have been created to noninvasively detect the level of glucose [56]. Both internal and external soft tissue, skin can be used as a mounting surface for wearable technology. The electrochemical detection of glucose in human blood samples is made possible by the immobilization of glucose oxidase and HRP in a Ca-Alginate hydrogel and its deposition onto an electrode surface. TMB is reduced in the presence of HRP using the hydrogen peroxide generated during the glucose oxidation process [57]. In clinical diagnostic-based biosensors, Ca-Alginate, poly (hydroxyethyl methacrylate) [poly (HEMA)], and polypyrrole (PPy) find conceivable uses. Glucose-responsive wearable insulin patch for blood glucose management is also developed using mechanisms like geometric alterations. The use of glucose directly for biosensing is still in the developmental stages [55]. This adaptable, patch acts as a standalone electrical device. Due to its biodegradability, it may even be used as a platform for an "implant and forget" sensor [58] if implanted.

Magnetic hydrogel-based neural stimulation: In order to design magnetic-responsive hydrogels, a hydrogel network is combined with magnetic-responsive fillers. Magnetic inclusions make normally nonresponsive hydrogels receptive to magnetic stimuli, allowing them to move, stretch, and change under the control of magnetic fields in a distant, controllable way [59].

Recent studies on magnetic hydrogels have revealed their role in the neural stimulation process. A. Tay et al. demonstrated that under the influence of a magnetic field magnetic hyaluronic hydrogel had regulatory effects on the growth of functional neurites and expression of inhibitory and excitatory ion channels. It was also shown that prolonged exposure of the magnetic hydrogel to magnetomechanical variation induced negative effects on the expression of ion channels responsible for causing pain [60]. Further, Y. Xu et al. showed that magnetic collagen hydrogel had the potential to guide the aligned growth of human tendon stem cells and promote their differentiation [61]. These findings open new doors to smart hydrogels in the tissue engineering domain.

Magnetic hyperthermia therapy: Hyperthermia is an upcoming approach for the treatment of cancer. Hyperthermia involves the heating of cells to a temperature of 42–46°C, which causes the denaturation of DNA and proteins [51], and eventually leads to the killing of the tumor cell [62]. Due to their accurate targeting and temperature controllability, magnetic nanoparticles (MNPs) have gained a lot of attention in the study of magneto-induced hyperthermia. MNPs are immobilized into the network of magnetic hydrogel to overcome the limitations of the conventional MNPs like low retention time and rapid degradation. It has been shown that the conduction of MNPs in magnetic hydrogels under a high-frequency alternating magnetic field causes enough thermogenesis to destroy tumor cells [63, 64].

Immuno-isolation: Immuno-isolation is established to shield the foreign material from the immune response of the host. Islet cells are delivered through injectable hydrogels in the treatment of diabetes type 1. Before being transplanted into diabetic recipients, islet cells are enclosed in the hydrogel, which creates a sufficient immune-isolation barrier to reduce rejection. 4-arm polyethylene glycol with maleimides makes a great choice for immuno-isolation applications. They are readily modified with thiolated molecules which allows regulation over the islet environment inside the host [15, 65].

Injectable hydrogels in cardiac tissue engineering: Although patch-based and cell sheet technologies have received a lot of research attention and show promising outcomes in cardiac tissue engineering, they call for a more intrusive surgical procedure. Currently, injectable hydrogels are of attention in cardiac tissue engineering as they are the least invasive and potential therapeutic agent that provides mechanical strength to the injured tissue [66]. Clinical trial results reveal that acellular alginate-based injectable hydrogels show considerable preservation of left ventricular indices and left ventricular ejection fraction [67] in patients with myocardial infarction. Extensive study is required in the domain of injectable hydrogels to make the therapeutic strategy stable (**Figure 1**).

Cardiac tissue regeneration: Studies on cardiac tissue regeneration are hindered by the complexity of cardiac cells [68]. Nevertheless, hydrogel-based studies have allowed researchers to get around some of these challenges. Cardiomyocytes derived from pluripotent stem cells were seeded into a collagen-fibrin hydrogel blend for the regeneration of cardiac tissue. It not only showed successful regeneration capacity but also showed that the regeneration of cardiac tissue was greatly influenced by the amount of cell seeding as well as the collagen-fibrin content [69].

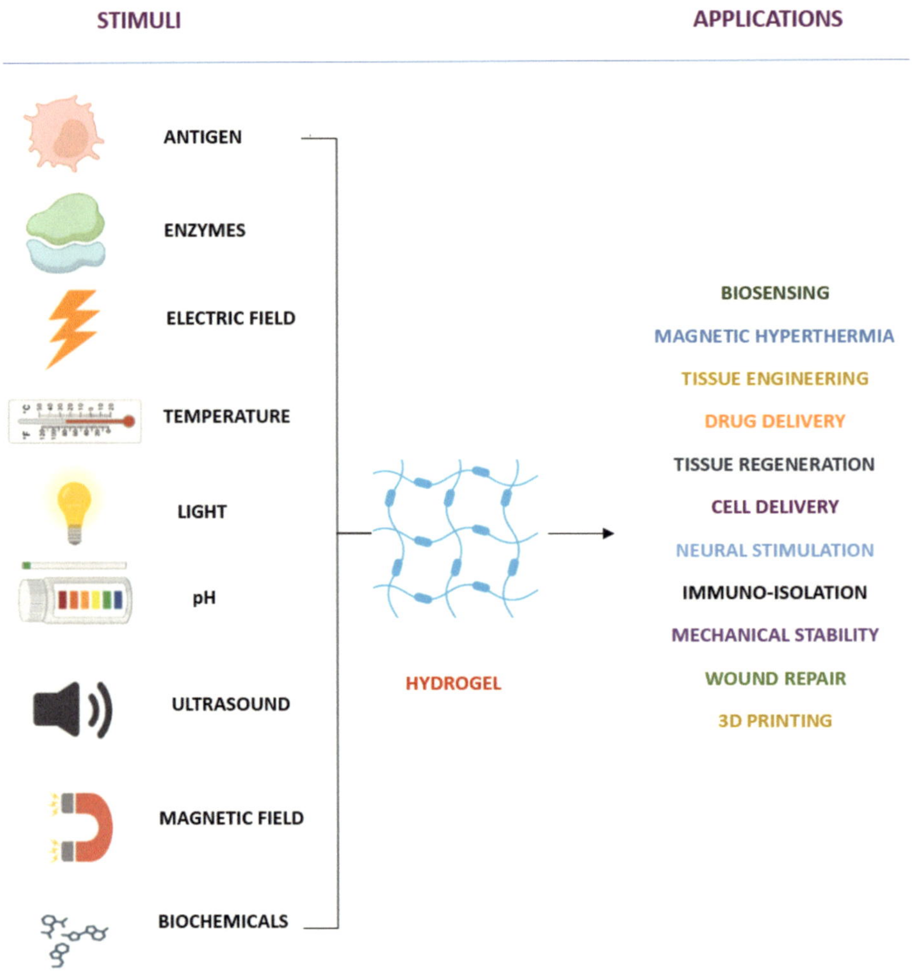

Figure 1.
Smart hydrogel: Stimuli and applications.

5. Factors affecting surface activity of the hydrogel

A hydrogel's practicality is greatly influenced by its mechanical behavior. Reduced mechanical resistance greatly hinders its optimal activity in biomedical applications, including tissue engineering. To obtain desired mechanical properties, synthetic polymers are preferred over natural polymers due to the ability to tailor them in accordance with their application. Crystalline, chemically [70, 71] crosslinked polymers can provide better mechanical stability due to their regularly arranged structure and easier to control nature. Moreover, the synthesis and use of chemically crosslinked polymer is pH independent, making its application easier [72].

Electrical conductivity of the polymer is another crucial factor required in specific applications. The use of electroactive biomaterials as a scaffold for tissue regeneration is typically considered. Recent studies have demonstrated how controlling cell adhesion, proliferation, movement, cell death, and differentiation with electroactive biomaterials can improve the regeneration of the heart, nerve, and bone [73]. Hydrogels utilized as substrates for electroactive tissue engineering must have conductivity levels that are comparable to those of biological tissues, the lack of which will potentially affect its application. There are several ways by which the conductance of a scaffolding material can be increased, such as

a. Boosting the ionic conductance of the aqueous phase: Preparing or swelling the hydrogel in an ion-rich aqueous solution will increase the ionic conductivity of the aqueous phase [74].

b. Inclusion of a conducting material in the hydrogel network: Through the dispersion of conductive nanomaterials or other conductive materials, such as carbon fibers, in the hydrogel framework, the integration of conductive materials attempts to produce a continuous network of electronic conduction throughout the hydrogel.

c. Employing a conducting polymeric unit in the gel matrix: Organic macromolecules having inherent electrical conductivity are known as conducting polymers. The conducting polymers polypyrrole (PPy), polyaniline (PANI), and poly-(3,4-ethylenedioxythiophene) (PEDOT) are frequently utilized to make such hydrogels [75].

Hydrogel is used for drug delivery owing to its qualities, including minimal cytotoxicity, biodegradability, and biocompatibility. Despite the fact that it has many benefits, it also has certain drawbacks. On contacting the drug-releasing medium, drug-carrying hydrogels typically exhibit a sudden release of drugs because of the presence of substantial proportion of water [76]. The incorporation of weakly soluble drugs, which are swiftly released through diffusion, is another issue with hydrogel's hydrophilic nature. These issues are effectively dealt with the incorporation of the drug into a nanomaterial, which is further encapsulated into a hydrogel matrix [37, 38].

6. Limitations of hydrogels

1. Hydrophobic therapeutical drugs have restricted loading quantity and uniformity in hydrogel matrices, and the affinity of hydrogels toward hydrophobic compounds is limited.

2. Natural hydrogels have poor mechanical properties, low reproducibility, and high production cost.

3. During delivery of the drug, hydrogels have complete dependence on swellability, diffusability of water, high chance of spontaneous rupture of drug during initial stages of hydrogel swelling, slow responsive time of stimuli-sensitive hydrogels with nonspecific drug release, fast dissolution, and uncontrollable porosity.

4. Hydrogels cannot be used in healing wet wounds as they cannot absorb large amounts of fluids.

5. Hydrogels are biodegradable but they sometimes release toxic exudates, which may not be able to be excreted *via* kidney and remain accumulated in the body.

6. Due to poor mechanical strength and low stability, it's difficult for hydrogels to store active substances.

7. Restriction in spatial manipulations of hydrogel, hence unable to attain full control on the organization and interaction among the cells, thus overall, there is no assurance on tissue morphogenesis.

7. Conclusion

Hydrogels differ from other forms of biomaterials in that they have a high-water content, a controlled swelling behavior, simple to handle, and are biocompatible, all of which make them desirable for biomedical applications. Hydrogels can respond to a variety of stimuli, including heat, pH, light, and chemical stimuli, based on their chemical structure and crosslink network, which can satisfy a variety of application needs. We provided detailed examples of both natural hydrogels made from polysaccharides and polypeptides as well as synthetic hydrogels, as well as different trends in their applications in various biomedical fields. It is important to conduct further research to develop materials with high mechanical strength, quick and effective self-healing, and a variety of biological activities for various specific limitations in biomedical applications. With extensive research trials and computational analysis, the employment of hydrogels can be effectively extended for various translational studies.

Author details

Ruby Varghese[1], Yogesh Bharat Dalvi[2*], P. Lochana[1], S. Achinthya[1], Bhagyashri Omprakash Somani[3], Preetha Karnaver[4], Nebu George Thomas[2,5], S. Rupesh[6], Nibu Varghese[2,7] and Jayachandran V.P.[8]

1 Department of Chemistry, School of Sciences, Jain Deemed to be University, Bengaluru, Karnataka, India

2 Pushpagiri Research Center, Pushpagiri Institute of Medical Sciences and Research Centre, Pushpagiri, Medical Society, Thiruvalla, Kerala, India

3 Department of Botany, Fergusson College (Autonomous), Pune, Maharashtra, India

4 Department of Zoology, Christian College Chengannur, Alappuzha, Kerala, India

5 Department of Periodontology, Pushpagiri College of Dental Sciences, Medicity, Pushpagiri Medical, Society, Thiruvalla, Kerala, India

6 Department of Pediatric and Preventive Dentistry, Malabar Dental College and Research Centre, Malappuram, Kerala, India

7 Mar Athanasios College for Advanced Studies Tiruvalla (MACFAST), Pathanamthitta, Kerala, India

8 Applied Biology Section, Applied Sciences Department, University of Technology and Applied Sciences, Al-Khuwair, Muscat, Sultanate of Oman

*Address all correspondence to: yogesh.botany@gmail.com

References

[1] Shmeis R, MA. Nanotechnology in Wastewater Treatment, Comprehensive Analytical Chemistry. Netherlands: Elsevier; 2022. pp. 105-134

[2] Peppas NA, Hoffman AS. 1.3.2E—hydrogels. In: Biomaterials Science. Cambridge: Academic Press; 2020

[3] Jo H, Yoon M, Gajendiran M, Kim K. Recent strategies in fabrication of gradient hydrogels for tissue engineering applications. Macromolecular Bioscience. 2020;**20**(3):1-11

[4] Calori IR, Braga G, Carvalho C, Bi H, Tedesco AC. Polymer scaffolds as drug delivery systems. European Polymer Journal. 2020;**129**:109621

[5] Jabbari E. Hydrogels for cell delivery. Gels. 2018;**4**(3):58

[6] Borges C, Colaço R, Serro AP. Poly (vinyl alcohol)-based hydrogels for joint prosthesis. Annals of Medicine. 2019;**51**(sup1):105

[7] Catoira MC, Fusaro L, Di Francesco D, Ramella M, Boccafoschi F. Overview of natural hydrogels for regenerative medicine applications. Journal of Materials Science: Materials in Medicine. 2019;**30**:1-10

[8] Soni V, Pandey V, Asati S, Gour V, Tekade RK. Biodegradable block copolymers and their applications for drug delivery. In: Basic Fundamentals of Drug Delivery. United States: Academic Press; 2019. pp. 401-447. DOI: 10.1016/b978-0-12-817909-3.00011-x

[9] Tang G, Zhou B, Li F, Wang W, Liu Y, Wang X, et al. Advances of naturally derived and synthetic hydrogels for intervertebral disk regeneration. Frontiers in Bioengineering and Biotechnology. 2020b;**8**:745

[10] Tang J, Katashima T, Li X, Mitsukami Y, Yokoyama Y, Sakumichi N, et al. Swelling behaviors of hydrogels with alternating neutral/highly charged sequences. Macromolecules. 2020a;**53**(19):8244-8254

[11] Yu T, Wang H, Zhang Y, Wang X, Han B. The delivery of RNA-interference therapies based on engineered hydrogels for bone tissue regeneration. Frontiers in Bioengineering and Biotechnology. 2020;**8**:445

[12] Madduma-Bandarage US, Madihally SV. Synthetic hydrogels: Synthesis, novel trends, and applications. Journal of Applied Polymer Science. 2021;**138**(19):50376

[13] He J, Burgess DJ. Impact of biomaterials' physical properties on cellular and molecular responses. In: Handbook of Biomaterials Biocompatibility. Duxford, England: Woodhead Publishing; 2020. pp. 69-84

[14] Cui L, Yao Y, Yim EK. The effects of surface topography modification on hydrogel properties. APL Bioengineering. 2021;**5**(3):031509

[15] Mantha S, Pillai S, Khayambashi P, Upadhyay A, Zhang Y, Tao O, et al. Smart hydrogels in tissue engineering and regenerative medicine. Materials. 2019;**12**(20):3323

[16] Ben Ammar NE, Barbouche M, Hamzaoui AH. In: Chen Y, editor. Hydrogels Based on Natural Polymers. Cambridge, MA, USA: Elsevier; 2020. p. 15

[17] Bai X, Gao M, Syed S, Zhuang J, Xu X, Zhang XQ. Bioactive hydrogels for

bone regeneration. Bioactive Materials. 2018;**3**(4):401-417

[18] Mariani E, Lisignoli G, Borzì RM, Pulsatelli L. Biomaterials: Foreign bodies or tuners for the immune response? International Journal of Molecular Sciences. 2019;**20**(3):636

[19] Palencia M, Lerma TA, Garcés V, Mora MA, Martínez JM, Palencia SL. Chapter 9 - Eco-friendly hydrogels. In: Palencia M, Lerma TA, Garcés V, Mora MA, Martínez JM, Palencia SL, editors. Advances in Green and Sustainable Chemistry. Eco-friendly Functional Polymers: Elsevier; 2021. pp. 141-153

[20] Egorikhina MN, Timofeeva LB, Linkova DD, Rubtsova YP, Bugrova ML, Charykova IN, et al. Biocompatibility study of hydrogel biopolymer scaffold with encapsulated mesenchymal stem cells. Polymers. 2023;**15**(6):1337

[21] Sánchez-Cid P, Jiménez-Rosado M, Rubio-Valle JF, Romero A, Ostos FJ, Rafii-El-Idrissi Benhnia M, et al. Biocompatible and thermoresistant hydrogels based on collagen and chitosan. Polymers. 2022;**14**(2):272

[22] Lan W, Xu M, Qin M, Cheng Y, Zhao Y, Huang D, et al. Physicochemical properties and biocompatibility of the bi-layer polyvinyl alcohol-based hydrogel for osteochondral tissue engineering. Materials & Design. 2021;**204**:109652

[23] Axpe E, Chan D, Offeddu GS, Chang Y, Merida D, Hernandez HL, et al. A multiscale model for solute diffusion in hydrogels. Macromolecules. 2019;**52**(18):6889-6897

[24] Nicolella P, Koziol MF, Löser L, Saalwächter K, Ahmadi M, Seiffert S. Defect-controlled softness, diffusive permeability, and mesh-topology of metallo-supramolecular hydrogels. Soft Matter. 2022;**18**(5):1071-1081

[25] Nicolella P, Lauxen D, Ahmadi M, Seiffert S. Reversible hydrogels with switchable diffusive permeability. Macromolecular Chemistry and Physics. 2021;**222**(16):2100076

[26] Vinchhi P, Rawal SU, Patel MM. Chapter 19 - Biodegradable hydrogels. In: Chappel E, editor. Developments in Biomedical Engineering and Bioelectronics, Drug Delivery Devices and Therapeutic Systems. United States: Academic Press; 2021. pp. 395-419

[27] Zong H, Wang B, Li G, Yan S, Zhang K, Shou Y, et al. Biodegradable high-strength hydrogels with injectable performance based on poly (l-glutamic acid) and gellan gum. ACS Biomaterials Science & Engineering. 2020;**6**(8):4702-4713

[28] Vázquez-González M, Willner I. Stimuli-responsive biomolecule-based hydrogels and their applications. Angewandte Chemie International Edition. 2020;**59**(36):15342-15377

[29] Li Z, Zhou Y, Li T, Zhang J, Tian H. Stimuli-responsive hydrogels: Fabrication and biomedical applications. Viewpoints. 2022;**3**(2):20200112

[30] Li X, Sun Q, Li Q, Kawazoe N, Chen G. Functional hydrogels with Tunable structures and properties for tissue engineering applications. Frontiers in Chemistry. 2018;**6**:1-20

[31] Chaudhary S, Chakraborty E. Hydrogel based tissue engineering and its future applications in personalized disease modeling and regenerative therapy. Beni-Suef University Journal of Basic and Applied Sciences. 2022;**11**:3

[32] Komatsu K, Shibata T, Shimada A, Ideno H, Nakashima K, Tabata Y, et al.

Cationized gelatin hydrogels mixed with plasmid DNA induce stronger and more sustained gene expression than atelocollagen at calvarial bone defects in vivo. Journal of Biomaterials Science, Polymer Edition. 2016;**27**(5):419-430

[33] Deev R, Plaksa I, Bozo I, Isaev A. Results of an international postmarketing surveillance study of pl-VEGF165 safety and efficacy in 210 patients with peripheral arterial disease. American Journal of Cardiovascular Drugs. 2017;**17**:235-242

[34] Carballo-Pedrares N, Fuentes-Boquete I, Díaz-Prado S, Rey-Rico A. Hydrogel-based localized nonviral gene delivery in regenerative medicine approaches - An overview. Pharmaceutics. 2020;**12**(8):752

[35] Chakradhar S. Treatments that made headlines in 2018. Nature Medicine. 2018;**24**(12):1785-1787

[36] Seidlits SK, Gower RM, Shepard JA, Shea LD. Hydrogels for lentiviral gene delivery. Expert Opinion on Drug Delivery. 2013;**10**(4):499-509

[37] Jiang Y, Krishnan N, Heo J, Fang RH, Zhang L. Nanoparticle–hydrogel superstructures for biomedical applications. Journal of Controlled Release. 2020;**324**:505-521

[38] Jiang Y, Wang Y, Li Q, Yu C, Chu W. Natural polymer-based stimuli-responsive hydrogels. Current Medicinal Chemistry. 2020c;**27**(16):2631-2657

[39] Nunes D, Andrade S, Ramalho MJ, Loureiro JA, Pereira MC. Polymeric nanoparticles-loaded hydrogels for biomedical applications: A systematic review on In vivo findings. Polymers (Basel). 2022;**14**(5):1010

[40] Mellati A, Hasanzadeh E, Gholipourmalekabadi M, Enderami SE. Injectable nanocomposite hydrogels as an emerging platform for biomedical applications: A review. Materials Science and Engineering: C. 2021;**131**:112489

[41] Sun Z, Wang X, Liu J, Wang Z, Wang W, Kong D, et al. ICG/l-arginine encapsulated PLGA nanoparticle-thermosensitive hydrogel hybrid delivery system for Cascade cancer photodynamic-NO therapy with promoted collagen depletion in tumor tissues. Molecular Pharmaceutics. 2021;**18**:928-939

[42] Wang Y, Li Q, Zhou J-E, Tan J, Li M, Xu N, et al. A photopolymerized semi-interpenetrating polymer networks-based hydrogel incorporated with nanoparticle for local chemotherapy of Tumors. Pharmaceutical Research. 2021;**38**:669-680

[43] Gao L, Yi M, Xing M, Li H, Zhou Y, Xu Q, et al. In situ activated mesenchymal stem cell (MSCs) by bioactive hydrogel for myocardial infarction treatment. Journal of Materials Chemistry B. 2020. DOI: 10.1039/d0tb01320j

[44] Yang M, He S, Su Z, Yang Z, Liang X, Wu Y. Thermosensitive injectable chitosan/collagen/β-glycerophosphate composite hydrogels for enhancing wound healing by encapsulating mesenchymal stem cell spheroids. ACS Omega. 2020;**5**(33):21015-21023

[45] Tony Thomas C, Ramanarayanan V, Krishnan V. Hydrogels - A versatile material in dentistry. European Journal of Molecular & Clinical Medicine (EJMCM). 2019;**6**(1):59-65

[46] Liu L, Wu D, Tu H, Cao M, Li M, Peng L, et al. Applications of hydrogels in drug delivery for Oral and maxillofacial diseases. Gels. 2023;**9**(2):146

[47] Huang Y, Li X, Yang L. Hydrogel encapsulation: Taking the therapy of mesenchymal stem cells and their derived Secretome to the next level. Frontiers in Bioengineering and Biotechnology. 2022b;**10**:1-9

[48] Huang M, Huang Y, Liu H, Tang Z, Chen Y, Huang Z, et al. Hydrogels for the treatment of oral and maxillofacial diseases: Current research, challenges, and future directions. Biomaterials Science. 8 Nov 2022;**10**(22):6413-6446

[49] Zagórska-Dziok M, Sobczak M. Hydrogel-based active substance release systems for cosmetology and dermatology application: A review. Pharmaceutics. 2020;**12**(5):396

[50] Chen M, Wang Y, Zhang J, Peng Y, Li S, Han D, et al. Stimuli-responsive DNA-based hydrogels for biosensing applications. Journal of Nanobiotechnology. 2022;**20**(1):1-22

[51] Zhang Y, Huang Y. Rational design of smart hydrogels for biomedical applications. Frontiers in Chemistry. 2021;**8**:615665

[52] Zhang Y, Dong L, Liu L, Wu Z, Pan D, Liu L. Recent advances of stimuli-responsive polysaccharide hydrogels in delivery systems: A review. Journal of Agricultural and Food Chemistry. 2022;**70**(21):6300-6316

[53] Bordbar-Khiabani A, Gasik M. Smart hydrogels for advanced drug delivery systems. International Journal of Molecular Sciences. 2022;**23**(7):3665

[54] Zhao Q, Li C, Shum HC, Du X. Shape-adaptable biodevices for wearable and implantable applications. Lab on a Chip. 2020;**20**(23):4321-4341

[55] Ying B, Liu X. Skin-like hydrogel devices for wearable sensing, soft robotics and beyond. Iscience. 2021;**24**(11):103174

[56] Wang J, Wang Z, Yu J, Kahkoska AR, Buse JB, Gu Z. Glucose-responsive insulin and delivery systems: Innovation and translation. Advanced Materials. 2020;**32**(13):1902004

[57] Roquero DM, Katz E. "Smart" alginate hydrogels in biosensing, bioactuation and biocomputing: State-of-the-art and perspectives. Sensors and Actuators Reports. 2022;**4**:100095

[58] Herrmann A, Haag R, Schedler U. Hydrogels and their role in biosensing applications. Advanced Healthcare Materials. 2021;**10**(11):2100062

[59] Li Z, Li Y, Chen C, Cheng Y. Magnetic-responsive hydrogels: From strategic design to biomedical applications. Journal of Controlled Release. 2021;**335**:541-556

[60] Tay A, Sohrabi A, Poole K, Seidlits S, Di Carlo D. A 3D magnetic hyaluronic acid hydrogel for magnetomechanical neuromodulation of primary dorsal root ganglion neurons. Advanced Materials. 2018;**30**(29):1800927

[61] Xu Y, Yin H, Chu J, Eglin D, Serra T, Docheva D. An anisotropic nanocomposite hydrogel guides aligned orientation and enhances tenogenesis of human tendon stem/progenitor cells. Biomaterials Science. 2021;**9**(4):1237-1245

[62] Wu J, Ning P, Gao R, Feng Q, Shen Y, Zhang Y, et al. Programmable ROS-mediated cancer therapy via magneto-inductions. Advanced Science. 2020;7(12):1902933

[63] Liu X, Zhang Y, Wang Y, Zhu W, Li G, Ma X, et al. Comprehensive understanding of magnetic hyperthermia

for improving antitumor therapeutic efficacy. Theranostics. 2020;**10**(8):3793

[64] Farzin A, Etesami SA, Quint J, Memic A, Tamayol A. Magnetic nanoparticles in cancer therapy and diagnosis. Advanced Healthcare Materials. 2020;**9**(9):1901058

[65] de Vos P. Bioartificial pancreas: Challenges and progress. Principles of Tissue Engineering. 2020:665-679

[66] Alonso JM, Andrade del Olmo J, Perez Gonzalez R, Saez-Martinez V. Injectable hydrogels: From laboratory to industrialization. Polymers. 2021;**13**(4):650

[67] Peña B, Laughter M, Jett S, Rowland TJ, Taylor MR, Mestroni L, et al. Injectable hydrogels for cardiac tissue engineering. Macromolecular Bioscience. 2018;**18**(6):1800079

[68] Hortensius RA, Lin WH, Ogle BM. Cardiac tissue engineering: A pathway for repair. In: Engineering in Medicine. United States: Academic Press; 2019. pp. 3-33

[69] Kaiser NJ, Kant RJ, Minor AJ, Coulombe KL. Optimizing blended collagen-fibrin hydrogels for cardiac tissue engineering with human iPSC-derived cardiomyocytes. ACS Biomaterials Science & Engineering. 2018;**5**(2):887-899

[70] You Y, Xing R, Zou Q, Shi F, Yan X. High-tolerance crystalline hydrogels formed from self-assembling cyclic dipeptide. Beilstein Journal of Nanotechnology. 2019;**10**:1894-1901

[71] Revete A, Aparicio A, Cisterna BA, Revete J, Luis L, Ibarra E, et al. Advancements in the use of hydrogels for regenerative medicine: Properties and biomedical applications. International Journal of Biomaterials. 2022:3606765

[72] Ye D, Chang C, Zhang L. High-strength and tough cellulose hydrogels chemically dual cross-linked by using low- and high-molecular-weight cross-linkers. Biomacromolecules. 2019;**20**(5):1989-1995

[73] Ning C, Zhou Z, Tan G, Zhu Y, Mao C. Electroactive polymers for tissue regeneration: Developments and perspectives. Progress in Polymer Science. 2018:144-162

[74] Peng Q, Chen J, Wang T, Peng X, Liu J, Wang X, et al. Recent advances in designing conductive hydrogels for flexible electronics. InfoMat. 2020;**2**. DOI: 10.1002/inf2.12113

[75] Kougkolos G, Golzio M, Laudebat L, Valdez-Nava Z, Flahaut E. Hydrogels with electrically conductive nanomaterials for biomedical applications. Journal of Materials Chemistry B. The Royal Society of Chemistry. 2023;**11**(10):2036-2062

[76] Ren Y, Li X, Han B, Zhao N, Mu M, Wang C, et al. Improved anti-colorectal carcinomatosis effect of tannic acid co-loaded with oxaliplatin in nanoparticles encapsulated in thermosensitive hydrogel. European Journal of Pharmaceutical Sciences. 2018;**128**:279-289

Chapter 5

Hydrogels and Nanogels as a Promising Carrier for Drug Delivery

Mohammed Hussain AL-Mayahy and Hiba Imad Hameed

Abstract

Among the drug delivery systems, hydrogels and nanogels have shown a vital role because of their advantageous 3D crosslinked networks. They have the propensity to absorb water due to their hydrophilic groups, making them excellent superabsorbents that are water-insoluble. Nanogels are crosslinked nano-sized hydrogels (20–200 nm) with greater tissue permeation due to their smaller size. Hydrogels and nanogels demonstrate many advantages, including biocompatibility, hydrophilicity, controlled drug release, and smart drug delivery. They are regarded as an interesting approach for the controlled release of medications since they can encapsulate drug molecules in their water-swollen network. Recent advances in polymer chemistry and nanotechnology have resulted in several significant improvements in the field of hydrogels and nanogels as drug delivery systems. In this chapter, the properties of hydrogels and nanogels, as well as their classification, drug release mechanisms, and applications for drug delivery, will be discussed.

Keywords: hydrogels, nanogels, drug delivery systems, drug release, tissue permeability

1. Introduction

Hydrogels are hydrophilic polymers with crosslinks that create a polymeric network. This enables them to absorb water at rates ranging from 10% to thousands of times their body weight [1, 2]. The presence of hydrophilic groups in the polymers forming hydrogel structures is attributed to their ability to absorb water. As a consequence of the contributions of various network's domains and groups, the hydration of the polymer will be at varying extent (often higher than 90% weight) based on the nature of the polymer and the aqueous environment. Because of the crucial crosslinks that are present in the hydrogel structure, hydrogels exhibit swelling rather than dissolving in the aqueous environment [3]. The term "crosslinker" refers to substances that join molecules and enhance the characteristics of hydrogels. Crosslinking hydrogels to create a three-dimensional structure increases molecular weight, offers better stability, and impacts physical qualities including polymer elasticity, viscosity, and polymer insolubility [4]. Alginate, chitosan, hyaluronic acid, and other natural biomaterials, as well as synthetic compounds like polyvinyl alcohol, polyacrylamide, and polyethylene glycol, can be used to prepare hydrogels [5].

Hydrogels are promising, fashionable, and intelligent drug delivery systems that meet the demands of precisely directing medications to the target site and managing drug release. Environmental, hydrolytic, or enzymatic stimuli can frequently alter hydrogels for drug release at the desired location [6]. Hydrogels provide a number of benefits, including the potential for biocompatibility, hydrophilicity, controlled drug release, and intelligent drug delivery [7]. Therefore, scientists from a variety of disciplines were particularly interested in creating and improving these delivery methods [1, 2]. However, there are drawbacks to using them. The hydrophobicity of most medicines is the main drawback of drug delivery. The hydrophilic polymeric core is practically not the best place to store lipophilic medications that are incompatible with it, which is a problem because many of the pharmaceuticals that are now utilised and successful in treating diseases are hydrophobic. Additionally, certain hydrogels have lower tensile strengths, which might lead to the medication being released early before it reaches the target location [8].

Major efforts have lately been made to use nanotechnology's potential in the delivery of various drugs. This provides a viable technique for site targeting and time-controlled delivery of different molecular weight drugs and bioactive chemicals. The nanotechnology approach is used to produce therapeutically useful compounds such as nanocapsules, nanoparticles, micellar systems, and conjugates. These preparations provide a multitude of advantages for drug administration, with the main advantages being improved medication safety and effectiveness. For instance, they can boost bioavailability, lengthen the time a drug or gene influences the target site, and improve drug stability against chemical or enzymatic degradation. They can also deliver drugs with more accuracy [3]. The hydrogel nanoparticulate materials (nanogels) would simultaneously exhibit the properties of both the nanoparticles and hydrogels that they individually possess. The hydrophilicity, adaptability, and biocompatibility of hydrogels as well as the benefits of nanoparticles, particularly their long half-life in blood and the potential to be targeted to the desired tissue, such as tumour sites, appear to benefit drug delivery [3].

2. Hydrogels

Hydrogels are comprised of 3D networks of hydrophilic polymers [9]. They have a large water absorption capacity due to the existence of hydrophilic functional groups connected to the 3D polymeric network of hydrogels [10]. The capacity to store water inside the network helps them to swell and collapse correctly, a characteristic that is helpful in medication delivery. Hydrogels are thought to be particularly promising materials for drug delivery because of their porosity and compatibility with aqueous environments [8]. Furthermore, the adjustable features of hydrogels make them ideal for certain medicinal applications such as oral drug delivery, ophthalmic, nasal, and transdermal routes [11].

2.1 Classification of hydrogels

There are several ways to categorise hydrogels, including:

1. Hydrogels either natural or synthetic. Natural hydrogels are a type of gel in which the polymers used to produce the gel are derived from natural sources.

The utilisation of natural polymers to make hydrogels has a variety of advantages, including biocompatibility, biodegradability, and non-toxicity. Natural polymers include polysaccharides and proteins. Synthetic hydrogels, on the other hand, are hydrogels created from synthetic polymers, such as polyacrylamide and its derivatives, polyvinyl alcohol (PVA), or polyethylene glycol (PEG), which are the building blocks of synthetic hydrogels. Synthetic polymers have recently superseded natural polymers in the production of hydrogels due to the several advantages they provide, such as longer shelf life, improved gel strength, and higher water absorption capacity [12].

2. The hydrogels' polymeric compositions (homo-, co-, and multi-polymer hydrogels). Homo-polymer hydrogels consist of just one kind of hydrophilic monomer; copolymer hydrogels, or network gels, consist of two types of monomers; and multi-polymer hydrogels or interpenetrating polymeric networks consist of three types of hydrophilic monomers [13].

3. Hydrogels' physical structure:

    - Amorphous or non-crystalline
    - Semi-crystalline: A complicated blend of crystalline and amorphous phases
    - Crystalline

    Lamellae, which are chain-folded layers that constitute the crystallisation of polymers, can be clustered to create spherulites, which are spherical components. Polymers have crystalline phases when the lamellar chains are ordered in well-known patterns; they display amorphous phases when the lamellae are randomly distributed [14].

4. According to the hydrogels' crosslinking type (chemical or physical). Chemical crosslinking is formed by a covalent bond that occurs between polymers and molecules to generate a 3D network structure. The stability and mechanical characteristics of hydrogels are their fundamental benefits. On the other hand, physical hydrogels are formed by noncovalent interactions between linear molecules due to the electrostatic, hydrophobic, and hydrogen bonds among these molecules. While the chemically crosslinked hydrogels are irreversible, the physically crosslinked hydrogels are reversible [15].

5. According to the electrical charge. Based on the electric charges that are positioned across the crosslinked chains, hydrogels may be divided into three major categories: ionic, non-ionic, and ampholytic hydrogels. Ionic hydrogels can be further subdivided into cationic and anionic hydrogels based on the type of electric charge that is present on the polymeric chain. Cationic hydrogels are sensitive to pH and can interact with other anionic molecules to create complexes. These qualities of cationic hydrogels make them useful as virus-alternative vectors for gene delivery and treatment. The majority of cationic hydrogels, including si-RNA, were effectively produced by enzymatic degradation. These hydrogels work well in cancer stem cell therapy [16].

6. According to the hydrogels' biodegradability. Both biodegradable and non-biodegradable hydrogels exist. Nonbiodegradable hydrogels are hydrogels that are resistant to environmental changes and may maintain their structure and physicochemical characteristics over an extended period of time. Biodegradable hydrogels are more prevalent in natural hydrogels than nonbiodegradable hydrogels [15].

2.2 Mechanisms for hydrogel matrix release

2.2.1 Diffusion-based release

The most typical way that drugs are released from hydrogels is by diffusion-based release [17]. The hydrogel matrix's mesh size has a significant impact on how quickly drugs diffuse out of it [18]. This is influenced by a number of factors, principally the extent of crosslinking, the monomer composition, and the nature and strength of environmental stimuli. Additionally, the size of the mesh has an important effect on the mechanical strength, degradability, diffusivity, and other physical characteristics of a hydrogel network [19, 20]. The usual mesh size of swollen hydrogels has been shown to range from 5 to 100 nm [20, 21], which is significantly bigger than the majority of small-molecule drugs. As a consequence, the diffusion of these medicines may only slightly slow in a swollen state. Macromolecules such as peptides and proteins, on the other hand, will have a prolonged release due to their hydrodynamic radii unless the structure and mesh size of the swollen hydrogels are suitably engineered to produce desired rates of macromolecular diffusion [22]. Drug release through a hydrogel mesh or water-filled hydrogel pores is made possible by diffusion-based drug delivery using either matrices or reservoirs. In the reservoir system, a core that contains the medicine is covered with a hydrogel membrane to create capsules or spheres with a high drug concentration in the system's centre, allowing for a time-independent and continuous drug release. The matrix system operates using the macromolecular mesh, demonstrating a time-dependent (not constant) drug release [8]. Both types of drug release are illustrated in **Figure 1**.

2.2.2 Swelling-based release

The hydrogels' swelling-based drug release technology is used for pharmaceuticals that are distributed inside of glassy polymers that swell when they come into contact with biofluids. Beyond its border, swelling causes expansion that facilitates drug diffusion and polymer chain relaxation. The procedure, also known as (Case II transport), enables consistent, time-independent release kinetics. The anomalous transfer is another term for this mechanism because it mixes the procedures of swelling and diffusion to enable drug release. The active ingredient can diffuse from an area of higher concentration to a lower one due to the gradient between the drug that is distributed in the hydrogel and its surrounding environment [8].

2.2.3 Chemically based release

Chemically based release is influenced by chemical processes that occur inside the matrix of the gel. Hydrolytic or enzymatic breakdown of polymeric chains is

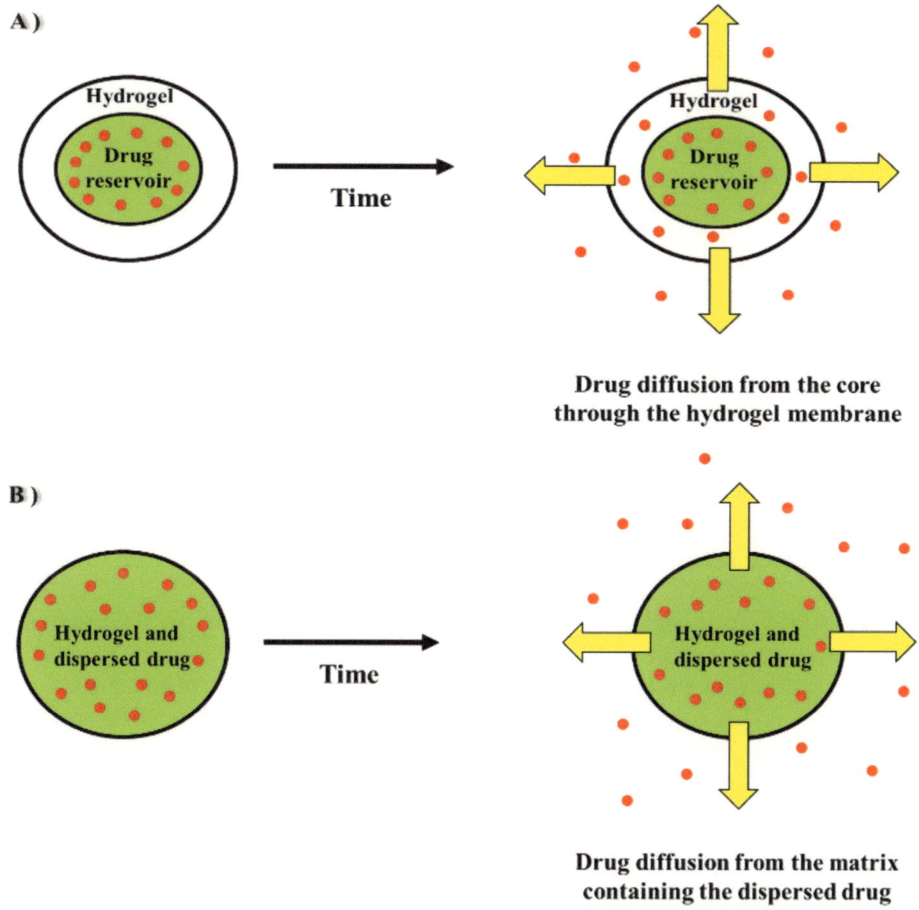

Figure 1.
(A): The core contains a drug in the Centre of the system in a higher concentration and is coated with a hydrogel membrane to produce a constant release rate in the reservoir system (time-independent), (B): Through the use of matrix delivery, the medicine is dissolved or dispersed throughout the hydrogel's three-dimensional structure (time-dependent), modified from [23].

one of these mechanisms, as well as reversible/irreversible interactions between the released medication and the polymeric network [3]. In chemically based delivery systems, drug release might occur by the breakage of polymeric chains via bulk or surface erosion, followed by the release of the entrapped drug from the hydrogels. The rate-limiting stage of chemically based release systems is polymer chain cleavage [17].

2.3 Controlled-delivery systems of hydrogel

Controlled-delivery systems are created to administer medication into the body in a certain, preset temporal and/or spatial manner to satisfy a particular therapeutic requirement. The hydrogels have special qualities that may make them one of the best controlled-release systems in the future among the various controlled-release systems that have been employed up to this point [3].

The controlled-release hydrogel systems are divided into two primary groups [3]:

1. Time-controlled systems

2. Stimulus-induced release systems (SIRS) [24, 25] are also known as stimulus-sensitive, stimulus-responsive, environmental-sensitive, environmental-responsive, or responsive hydrogel systems.

Responsive hydrogels drastically modify their structure and behaviour in response to changes in the external environment [3]. The environment-sensitive hydrogel systems, often known as smart systems, can be divided into the following categories:

1. Physically triggered release mechanisms

2. Chemically triggered release mechanisms

3. Other stimulus-induced release mechanisms

In this context, physical stimuli of interest include those involving temperature, electric current, pressure, sound, light, and magnetic fields. In contrast, specific molecular recognition events, ions, solvent composition, pH, and others are chemical triggers [25, 26]. Due to their capacity for recurrent conversion of swelling to deswelling in response to external temperature variations, temperature-sensitive (thermo-responsive) hydrogels have attracted a lot of attention [27, 28]. Contrarily, hydrogel

	Limitation	**Solution**
1	Some hydrogels have nonbiodegradable and non-biocompatible characteristics.	Creation of hydrogels that are both biocompatible and biodegradable, such as PEG-PLGA-PEGa, or use of polymers with hydrolyzable moieties (chemical modification) [29].
2	Response times of stimulus-sensitive hydrogels are too long.	Creating quick-acting hydrogels that are thinner and smaller [7].
3	Fast drug release from large porous hydrogels and a quick burst release of medication during hydrogel swelling.	Prior to gelation, drugs can be physically or covalently tethered to the polymer chains (tethering approach) [30]. To combat rapid disintegration and quick drug release, covalent crosslinking is used to generate di-block or tri-block copolymers [6].
4	In the entrapment procedure, there is a chance that the medication will become inactive and release in a burst.	Use of the tethering technique [30].
5	possible drug inactivation during the covalent binding method's polymer binding.	Replacing direct covalent drug binding to the polymer with suitable linkers that allow for controlled drug release [17].
6	Hydrogel with diffusion-controlled release has an unspecific drug release mechanism.	Controlling the drug release from hydrogels using chemically and physiologically induced release triggers [30].

	Limitation	Solution
7	Possibility of toxicity of small-molecule crosslinkers that are left behind after a hydrogel is created using a small-molecule crosslinking approach.	Using the Michael addition process or the Schiff base synthesis approach to crosslink polymers to other polymers [6].
8	For oral drug delivery systems, chitosan-based hydrogel matrices quickly dissolve in stomach acid.	Crosslinking, conjugation, or the creation of polyelectrolyte complexes [17]. For the administration of oral insulin, scientists have created hydrogels based on chitosan-poly (g-glycolic acid) [31].
9	Drug-loaded colloidal carriers are difficult to incorporate into hydrogels.	Preparation of liposomes in hydrogels as a mixed delivery system [32].
10	Non-homogenous hydrophobic drug dispersion inside hydrogels and limited hydrophobic medication delivery using hydrogels.	Directly incorporating hydrophobic domains into the hydrogel network [17]. Creation of a weakly soluble drug's solid molecular dispersion [17]. Using polymer nanoparticles to encapsulate the medicine allows for a well-dispersed drug slurry to be produced [17]. Preparation of drug delivery system made of hydrogel and nanoparticles [6, 33].
11	Poloxamer-based thermosensitive hydrogels have low biodegradability and possible toxicity.	Preparation of biodegradable and biocompatible hydrophobic polyesters such as PLGA and PCLb that are crosslinked to PEG-based hydrogels [34]. A suitable in situ forming hydrogel that is temperature-sensitive is made of the PEG-PNIPAAMC copolymer. [34].
12	Response of light-responsive hydrogels to stimuli (light) is slow and variable.	Fabrication of 2-hydroxyethyl methacrylate crosslinked polymers with azobenzene functional groups [35]. The interaction of polyacrylamide and polyacrylic acid forms an interpenetrating polymer network [36].
13	Clogging of the needle during the injection of hydrogels that are temperature and pH sensitive.	pH/thermosensitive hydrogels are created using a dual-responsive hydrogel technique [17].
14	Solubility, pH-related activity, and the delayed sol-gel transition phase.	Chemical change to enhance mucoadhesiveness and solubility profile [37].
15	PEG hydrogels, as a tissue engineering scaffold, do not have any particular cell adhesive characteristics.	Using extracellular matrix proteins, PEG hydrogels may be modified to act as cell adhesives [38].
16	Low drug loading and quick drug release.	Incorporating N-(3-aminopropyl) methacrylamide or other ionic or hydrophobic monomers into pHEMAd hydrogels to adjust drug release rate and boost drug loading capacity [39].
17	the difficulty of surgically implanting pre-formed hydrogels or scaffolds used in tissue engineering, and the danger of infection.	To get around these scaffolding restrictions, use injectable hydrogel systems instead [40]. Hydrogels with micro-engineered structures are created [41].

	Limitation	Solution
18	Low hydrogel network loading capacity for DNA or RNA and restricted transgenic expression capacity for gene transfer.	DNA/polymer polyplexes are incorporated into PEG hydrogel scaffolds [42]. Using hydrogels made of fibrin and hyaluronic acid as scaffolding for DNA/polymer polyplexes [43].
19	The ineffectiveness and unsuitability as a carrier for hydrophobic and small-molecule active medicines.	The creation of hydrogel carriers made from copolymers of acrylic acid and methyl methacrylate serves as a unique oral drug delivery strategy for hydrophobic active medicines with tiny molecular weights [44].
20	Hydrogels made of calcium alginate have low mechanical strength.	Agar/alginate beads' composition, which has the advantages of improved mechanical strength and regulated drug release [45].

Table 1.
Limitations of hydrogels and possible solutions.

systems with chemical responsiveness suggest a number of forms of hydrogels in which medicine can be released from a repository in response to variations in the amount of a particular molecule or biologically active substance in the environment. Due to their distinctive pH-dependent swelling–deswelling behaviour, hydrogels with pH-responsive systems are extremely important [3].

The main problem with stimulus-sensitive hydrogels is how slowly they react. The easiest technique to produce fast-acting responsiveness is to create hydrogels that are thinner and smaller. However, this reduces the mechanical strength of the hydrogel device and the polymer network as a whole [7]. **Table 1** lists all the restrictions associated with hydrogels as well as potential remedies.

3. Hydrogel nanoparticles (nanogels)

Nanogels are submicron-sized, three-dimensional crosslinked polymer networks. Nanogel combines the properties of hydrogels and nanoparticles since it is made up of hydrogel particulate entities with nanometre-sized spaces [46]. Nanogels can be created through monomer-heterogeneous polymerisation or from precursors of polymers. Crosslinking, including physical and chemical crosslinking, is a crucial step in the fabrication of nanogels. Nanogels exhibit the swelling capability, or capacity, to absorb large volumes of water or biological fluids while retaining their structural integrity. Nanogels are an interesting prospect for several applications due to their special characteristics [3]. Nanogel has been found to have great drug loading capacity, biological consistency, higher stability, and sensitivity to a wide variety of environmental stimuli (such as pH, temperature, and ionic strength) in comparison to other nanocarriers [46].

Drugs loaded within nanogels can easily pass through physiological barriers due to their nanosize range, thus increasing drug bioavailability. In addition, with the use of nanogels, less medication is required, and there are fewer doses per day, which reduces the toxicity of the medication [47]. However, one should pay attention to the solvents and the type and concentration of surfactants used in the preparation of nanogels because toxicity due to surfactants can occur occasionally [47].

4. Hydrogels and nanogels applications in drug delivery

4.1 Oral drug delivery

The most popular, desired, and patient-compliant method of medication administration is oral dosing. Copolymer hydrogel networks are an appropriate drug delivery vehicle because they enhance oral absorption and bioavailability. As a mucoadhesive substance with the potential to prolong medication release and absorption, hydrogels are regarded as secure drug delivery systems for oral administration. The ability to prevent integrated drugs from enzymatic degradation is hydrogels' additional benefit as oral drug delivery vehicles [48]. In order to avoid the difficulties involved in administering insulin parenterally, hydrogels are primarily explored for oral medication delivery of insulin [31].

4.2 Parenteral drug delivery

The parenteral method of administration is the most preferred route of administration for several medications, including peptides and proteins. For parenteral medication administration, hydrogels may be employed as controlled drug delivery systems. Drugs can be protected against enzymatic degradation, their half-life is extended, their bioavailability is increased, their frequency of administration is decreased, and as a result, their compliance with the patient's needs is increased by using hydrogels. Temperature-sensitive injectable hydrogels are typically sol (fluid) at ambient temperature and gel (viscous) at human temperature. Drug release can be sustained, and medication bioavailability can be increased via gelation. The most popular temperature-sensitive hydrogels utilised in parenteral drug delivery systems are those based on poloxamer, although they are limited by the fact that they are not biodegradable [49].

4.3 Nasal drug delivery

High patient compliance and the avoidance of the hepatic first-pass effect, which might boost medication bioavailability, are two benefits of nasal drug administration. However, this method of administration also has its own drawbacks, such as the mucosal membranes acting as a barrier to macromolecule absorption and the short nasal residence period caused by mucosal turnover. Chitosan hydrogels, for example, are highly regarded as innovative nasal delivery systems that might lengthen the period that loaded active chemicals remain in the nasal cavity. These hydrogels have mucoadhesive, viscoelastic, and biocompatible qualities. The majority of the polymers used in nasal administration systems have a high thermal sensitivity and can gel at the site of action at body temperature [37].

4.4 Ocular drug delivery

A frequent method of topical drug administration for delivering medications to the eye is eye drops. Although this method is ineffective and might have systemic side effects, only around 5% of the incorporated drug would penetrate the intraocular tissue; the remaining 95% would be lost to tear drainage. Furthermore, the medication has a relatively brief residence duration in the eye. Because of this, it would be extremely desirable to create innovative drug delivery methods that would lengthen the period that medication is in the eye, reduce drug loss, and have fewer systemic negative effects [50]. Numerous studies have investigated contact lenses as ocular drug delivery devices, which have the benefit of extending the period that drugs are present in the body and their bioavailability. Because of their transparency and biocompatibility, hydrogel contact lenses have received a lot of attention as ocular medication delivery devices. Regarding this, drug molecules might be uniformly disseminated in matrices of hydrogels such as hydroxyethyl methacrylate (HEMA) polymer. However, this technique is only applicable to water-soluble medicines and could result in rapid drug release [50]. Hydrophobic or ionic monomers might be added to hydroxyethyl methacrylate (HEMA) hydrogels to promote hydrogel and drug interactions, regulate the rate of drug release, and boost loading capacity for drugs [39]. Dexamethasone-containing eye drops were created using either solvent evaporation or emulsification process with 2-hydroxypropyl-γ-cyclodextrin (HP-γ-CDs) medium containing cyclodextrin nanogel for sustained release. Using pH-sensitive polyvinylpyrrolidone-poly [acrylic acid] (PVP/PAAc) nanogels, pilocarpine was encapsulated to improve bioavailability, stability, and maintain a sufficient concentration of the medication at the site of action for a prolonged length of time [51].

4.5 Topical delivery

Delivering medications topically is a popular way of administration that is used to decrease side effects and localise large quantities of a medicine at the target site. Hydrogels are regarded as acceptable carriers for topical medication delivery due to their low toxicity potential and prolonged drug release. Hydrogels also offer the advantages of biocompatibility, softness, and high-water content, which can mirror the features of real tissues. Furthermore, due to hydrogels' swelling and moisturising properties, they might prevent irritation of the enclosing tissues. Another significant feature of hydrogels is their capacity to preserve drugs from severe environmental conditions [52]. Nanogels have been used in dermatology and cosmetics as topical delivery methods for nonsteroidal anti-inflammatory medications (NSAIDs), as well as to treat psoriatic plaque and allergic contact dermatitis. For this purpose, nanogels are appropriate because they address a significant constraint of topically delivering systems: the relatively limited contact period between active medicines and the site of application. Thus, a homogenous nanogel dispersion is produced, maintaining water in the gel matrix [51].

4.6 Transdermal and subcutaneous delivery

In comparison to other administration methods, the transdermal route avoids the first-pass effect, improves therapeutic efficacy, provides plasma drug concentrations at a steady level, and increases patient adherence. Transdermal nanogels can produce an increase in permeability and stability in comparison to conventional oral administration [53]. Swollen hydrogels were employed in wound dressings as a controlled-release technique. The anti-inflammatory medication budesonide has been transported via hydrogels. To improve nicotine and hormone product penetration, these hydrogels are being investigated for transdermal iontophoresis. The favourable biodegradability of hydrogels encourages the development of biodegradable implanted hydrogels. These are mostly employed in the treatment of cancer by delivering anticancer medicines subcutaneously. It is made from crosslinked poly hydroxyethyl methacrylate (PHEMA), also used to deliver the chemotherapeutic agent cytarabine [14].

4.7 Brain drug delivery

The blood-brain barrier is still a major issue in drug delivery to the brain, which is still a difficult process. Local medication delivery via brain implants is a possibility, but it comes with the risk of infection and harm to brain tissue. Epicortical delivery utilising hydrogels is a different strategy for brain local administration that might distribute the loaded medicine directly into the brain with little tissue damage [54].

4.8 Tissue engineering

For tissue engineering, hydrogels offer several benefits including resemblance to extracellular matrices of tissues, promotion of cell proliferation, minimal irritation to surrounding tissues, and prolonged release of integrated growth factors. Injectable hydrogels are superior to other traditional scaffolds in several ways, including simplicity of handling, deeper tissue penetration, greater margin adaptability, and less invasiveness [40]. Other interesting engineering techniques to address the problems that tissue engineering is now encountering are micro-engineered hydrogels [41].

4.9 Gene delivery

Delivering DNA or RNA for the purpose of genetic alteration can be achieved using a hydrogel scaffold. Hydrogels can improve the effectiveness of gene therapy, particularly in the treatment of cancer. In cancer therapy, siRNA or fatal genes would be enclosed in hydrogel scaffolds and would encourage malignant cells to undergo cell death. However, hydrogels used for gene delivery may have some drawbacks, such as a low capacity for loading genes and a quick release of encapsulated genes. To overcome these restrictions, a number of techniques have been proposed, such as the condensation of DNA or RNA in nanoparticulate systems and subsequent encapsulation in a hydrogel scaffold [43].

4.10 Cosmetics application

The primary function of the skin is to protect the body from harmful external environmental factors like pathogens and UV radiation. Additionally, it helps keep the body hydrated and at the proper temperature. To enhance the skin's appearance and texture, cosmetic products are frequently applied. Skin cleansers, moisturisers, and body lotions are all needed to maintain a healthy skin texture. Hydrogel-based cosmetic products are now gaining popularity in the cosmetic industry due to benefits such as biocompatibility, elasticity, softening, and increased water content. Wrinkles, cellulite, pigmentation, ageing, and skin hydration can all be efficiently treated using hydrogels. Caffeine-containing bioadhesive hydrogels, for example, are widely employed in cosmetic applications. The bioadhesive property of hydrogels aids in the gradual release of caffeine into the skin, improving skin texture and appearance [14].

4.11 Vaccines delivery

Immune responses that are specific to an antigen are induced during vaccination. Polymeric nanogels are a unique, alternative method of vaccine administration that is being used to improve the potency and effectiveness of vaccinations. Nanogels perform better than conventional immunisations in preventing the enzymatic breakdown of vaccine antigens. Vaccine administration with regard to a specified target can be considerably enhanced by using surface-altered nanogels with antibodies' affixes and other ligands [55].

4.12 Cancer therapy

Many anticancer drugs such as doxorubicin, cisplatin, temozolamide, and 5-FU have been loaded into nanogels for the treatment of different types of cancer. For example, nanogels loaded with chitin-polymerised doxorubicin are employed for the management of lung, breast, liver, and prostate cancer. In cancer treatment, drug delivery is required to be efficient and highly specific to the target site with low toxicity to the adjacent healthy tissue. Thermosensitive and pH-sensitive nanogels containing cisplatin were utilised for the management of breast cancer [56]. Likewise, fludarabine loaded into polyplex nanogels was employed to enhance the activity and reduce the toxicity of the drug [57].

5. Conclusion

Hydrogels and nanogels can be regarded as promising carriers for drug delivery since they have shown a positive role in the delivery of various types of drugs and bioactive molecules by different routes of administration. They demonstrated superiority in drug release, targeting, and stability in comparison to conventional dosage forms due to their unique features. However, to increase their commercialisation, some issues are required to be addressed, including the development of cost-effective manufacturing methods for large-scale production and the precise control of drug pharmacokinetics to reach the target site at an adequate dose. The developments in nanotechnology and polymer sciences may resolve these challenges, which make hydrogels and nanogels an excellent platform for drug delivery in clinical settings.

Conflict of interest

"The authors declare no conflict of interest."

Author details

Mohammed Hussain AL-Mayahy* and Hiba Imad Hameed
College of Pharmacy, Mustansiriyah University, Baghdad, Iraq

*Address all correspondence to: mohammedhussain@uomustansiriyah.edu.iq

References

[1] Hoffman AS. Hydrogels for biomedical applications. Advanced Drug Delivery Reviews. 2002;**54**(1):3-12. DOI: 10.1016/s0169-409x(01)00239-3

[2] Omidian H, Park K. Swelling agents and devices in oral drug delivery. Journal of Drug Delivery Science and Technology. 2008;**18**(2):83-93. DOI: 10.1016/S1773-2247(08)50016-5

[3] Hamidi M, Azadi A, Rafiei P. Hydrogel nanoparticles in drug delivery. Advanced Drug Delivery Reviews. 2008;**60**(15):1638-1649. DOI: 10.1016/j.addr.2008.08.002

[4] Husnaini Zainal S, Hanisah Mohd N, Suhaili N, Hannan Anuar F, Mat Lazim A, Othaman R. Preparation of cellulose-based hydrogel: a review. Journal of Materials Research and Technology. 2021;**10**:935-952. DOI: 10.1016/j.jmrt.2020.12.012

[5] Su J, Li J, Liang J, Zhang K, Li J. Hydrogel preparation methods and biomaterials for wound dressing. Life. 2021;**11**(10):1016. DOI: 10.3390/life11101016

[6] Hoare TR, Kohane DS. Hydrogels in drug delivery: Progress and challenges. Polymer. 2008;**49**(8):1993-2007. DOI: 10.1016/j.polymer.2008.01.027

[7] Qiu Y, Park K. Environment-sensitive hydrogels for drug delivery. Advanced Drug Delivery Reviews. 2001;**53**(3):321-339. DOI: 10.1016/s0169-409x(01)00203-4

[8] Narayanaswamy R, Torchilin VP. Hydrogels and their applications in targeted drug delivery. Molecules. 2019;**24**(3):603. DOI: 10.3390/molecules24030603

[9] Bahram M, Mohseni N, Moghtader M. An introduction to hydrogels and some recent applications. In: Majee SB, editor. Emerging Concepts in Analysis and Applications of Hydrogels. Croatia: InTech; 2016, pp. 9-38. DOI: 10.5772/64301

[10] Ahmed EM. Hydrogel: Preparation, characterization, and applications: A review. Journal of Advanced Research. 2015;**6**(2):105-121. DOI: 10.1016/j.jare.2013.07.006

[11] Simões S. Modular hydrogels for drug delivery. Journal of Biomaterials and Nanobiotechnology. 2012;**3**(2):185-199. DOI: 10.4236/jbnb.2012.32025

[12] Soni RK, Kewatkar SM, Jain V, Saluja MS. A review of hydrogels, with its properties and applications in medicine. Journal of Biomedical and Pharmaceutical Research. 2023;**12**(2):20-33. DOI: 10.32553/jbpr.v12i2.970

[13] Vashist A, Ahmad S. Hydrogels: smart materials for drug delivery. Oriental Journal of Chemistry. 2013;**29**(3):861-870. DOI: 10.13005/ojc/290303

[14] Kasai RD, Radhika D, Archana S, Shanavaz H, Koutavarapu R, Lee D-Y, et al. A review on hydrogels classification and recent developments in biomedical applications. International Journal of Polymeric Materials and Polymeric Biomaterials. 2022;**72**:1-11. DOI: 10.1080/00914037.2022.2075872

[15] Zheng H, Li M, Wu L, Liu W, Liu Y, Gao J, et al. Progress in the application of hydrogels in immunotherapy of gastrointestinal tumors. Drug Delivery. 2023;**30**(1):2161670. DOI: 10.1080/10717544.2022.2161670

[16] Kaith BS, Singh A, Sharma AK, Sud D. Hydrogels: Synthesis, classification, properties and potential applications—A brief review. Journal of Polymers and the Environment. 2021;**29**(12):3827-3841. DOI: 10.1007/s10924-021-02184-5

[17] Ghasemiyeh P, Mohammadi-Samani S. Hydrogels as drug delivery systems; pros and cons. Trends in Pharmaceutical Sciences. 2019;**5**(1):7-24. DOI: 10.30476/tips.2019.81604.1002

[18] Canal T, Peppas NA. Correlation between mesh size and equilibrium degree of swelling of polymeric networks. Journal of Biomedical Materials Research. 1989;**23**(10):1183-1193. DOI: 10.1002/jbm.820231007

[19] Amsden B. Solute diffusion within hydrogels. Mechanisms and Models. Macromolecules. 1998;**31**(23):8382-8395. DOI: 10.1021/ma980765f

[20] Mason MN, Metters AT, Bowman CN, Anseth KS. Predicting controlled-release behavior of degradable PLA-b-PEG-b-PLA hydrogels. Macromolecules. 2001;**34**(13):4630-4635. DOI: 10.1021/ma010025y

[21] Cruise GM, Scharp DS, Hubbell JA. Characterization of permeability and network structure of interfacially photopolymerized poly(ethylene glycol) diacrylate hydrogels. Biomaterials. 1998;**19**(14):1287-1294. DOI: 10.1016/S0142-9612(98)00025-8

[22] Lustig SR, Peppas NA. Solute diffusion in swollen membranes. IX. Scaling laws for solute diffusion in gels. Journal of Applied Polymer Science. 1988;**36**(4):735-747. DOI: 10.1002/app.1988.070360401

[23] Caló E, Khutoryanskiy VV. Biomedical applications of hydrogels: A review of patents and commercial products. European Polymer Journal. 2015;**65**:252-267. DOI: 10.1016/j.eurpolymj.2014.11.024

[24] Lin C-C, Metters AT. Hydrogels in controlled release formulations: Network design and mathematical modeling. Advanced Drug Delivery Reviews. 2006;**58**(12-13):1379-1408. DOI: 10.1016/j.addr.2006.09.004

[25] Kikuchi A, Okano T. Pulsatile drug release control using hydrogels. Advanced Drug Delivery Reviews. 2002;**54**(1):53-77. DOI: 10.1016/s0169-409x(01)00243-5

[26] Miyata T, Uragami T, Nakamae K. Biomolecule-sensitive hydrogels. Advanced Drug Delivery Reviews. 2002;**54**(1):79-98. DOI: 10.1016/s0169-409x(01)00241-1

[27] Bromberg LE, Ron ES. Temperature-responsive gels and thermogelling polymer matrices for protein and peptide delivery. Advanced Drug Delivery Reviews. 1998;**31**(3):197-221. DOI: 10.1016/s0169-409x(97)00121-x

[28] Dinarvand R, D'Emanuele A. The use of thermoresponsive hydrogels for on-off release of molecules. Journal of Controlled Release. 1995;**36**(3):221-227. DOI: 10.1016/0168-3659(95)00035-7

[29] Li Z, Ning W, Wang J, Choi A, Lee P-Y, Tyagi P, et al. Controlled gene delivery system based on thermosensitive biodegradable hydrogel. Pharmaceutical Research. 2003;**20**(6):884-888. DOI: 10.1023/A:1023887203111

[30] Bhattarai N, Gunn J, Zhang M. Chitosan-based hydrogels for controlled, localized drug delivery. Advanced Drug Delivery Reviews. 2010;**62**(1):83-99. DOI: 10.1016/j.addr.2009.07.019

[31] Chaturvedi K, Ganguly K, NadagoudaMN,AminabhaviTM.Polymeric hydrogels for oral insulin delivery. Journal of Controlled Release. 2013;**165**(2):129-138. DOI: 10.1016/j.jconrel.2012.11.005

[32] Hurler J, Berg OA, Skar M, Conradi AH, Johnsen PJ, Škalko-Basnet N. Improved burns therapy: Liposomes-in-hydrogel delivery system for Mupirocin. Journal of Pharmaceutical Sciences. 2012;**101**(10):3906-3915. DOI: 10.1002/jps.23260

[33] Gou M, Li X, Dai M, Gong C, Wang X, Xie Y, et al. A novel injectable local hydrophobic drug delivery system: Biodegradable nanoparticles in thermo-sensitive hydrogel. International Journal of Pharmaceutics. 2008;**359**(1-2):228-233. DOI: 10.1016/j.ijpharm.2008.03.023

[34] Alexander A, Ajazuddin K, Saraf JS, Saraf S. Poly(ethylene glycol)–poly(lactic-co-glycolic acid) based thermosensitive injectable hydrogels for biomedical applications. Journal of Controlled Release. 2013;**172**(3):715-729. DOI: 10.1016/j.jconrel.2013.10.006

[35] Unger K, Salzmann P, Masciullo C, Cecchini M, Koller G, Coclite AM. Novel light-responsive biocompatible hydrogels produced by initiated chemical vapor deposition. ACS Applied Materials & Interfaces. 2017;**9**(20):17408-17416. DOI: 10.1021/acsami.7b01527

[36] Xu Y, Ghag O, Reimann M, Sitterle P, Chatterjee P, Nofen E, et al. Development of visible-light responsive and mechanically enhanced 'smart' UCST interpenetrating network hydrogels. Soft Matter. 2018;**14**(1):151-160. DOI: 10.1039/c7sm01851g

[37] Nazar H, Fatouros DG, van der Merwe SM, Bouropoulos N, Avgouropoulos G, Tsibouklis J, et al. Thermosensitive hydrogels for nasal drug delivery: The formulation and characterisation of systems based on N-trimethyl chitosan chloride. European Journal of Pharmaceutics and Biopharmaceutics. 2011;**77**(2):225-232. DOI: 10.1016/j.ejpb.2010.11.022

[38] Zhu J. Bioactive modification of poly(ethylene glycol) hydrogels for tissue engineering. Biomaterials. 2010;**31**(17):4639-4656. DOI: 10.1016/j.biomaterials.2010.02.044

[39] Hu X, Hao L, Wang H, Yang X, Zhang G, Wang G, et al. Hydrogel contact lens for extended delivery of ophthalmic drugs. International Journal of Polymer Science. 2011;**2011**:e814163. DOI: 10.1155/2011/814163

[40] Sivashanmugam A, Arun Kumar R, Vishnu Priya M, Nair SV, Jayakumar R. An overview of injectable polymeric hydrogels for tissue engineering. European Polymer Journal. 2015;**72**:543-565. DOI: 10.1016/j.eurpolymj.2015.05.014

[41] Khademhosseini A, Langer R. Microengineered hydrogels for tissue engineering. Biomaterials. 2007;**28**(34):5087-5092. DOI: 10.1016/j.biomaterials.2007.07.021

[42] Lei Y, Huang S, Sharif-Kashani P, Chen Y, Kavehpour P, Segura T. Incorporation of active DNA/cationic polymer polyplexes into hydrogel scaffolds. Biomaterials. 2010;**31**(34):9106-9116. DOI: 10.1016/j.biomaterials.2010.08.016

[43] Lei Y, Rahim M, Ng Q, Segura T. Hyaluronic acid and fibrin hydrogels with concentrated DNA/PEI polyplexes for local gene delivery. Journal of Controlled Release. 2011;**153**(3):255-261. DOI: 10.1016/j.jconrel.2011.01.028

[44] Caldorera-Moore M, Maass K, Hegab R, Fletcher G, Peppas N. Hybrid responsive hydrogel carriers for oral delivery of low molecular weight therapeutic agents. Journal of Drug Delivery Science and Technology. 2015;**30**:352-359. DOI: 10.1016/j.jddst.2015.07.023

[45] Yin Z-C, Wang Y-L, Wang K. A pH-responsive composite hydrogel beads based on agar and alginate for oral drug delivery. Journal of Drug Delivery Science and Technology. 2018;**43**:12-18. DOI: 10.1016/j.jddst.2017.09.009

[46] Zhang H, Zhai Y, Wang J, Zhai G. New progress and prospects: The application of nanogel in drug delivery. Materials Science and Engineering:C. 2016;**60**:560-568. DOI: 10.1016/j.msec.2015.11.041

[47] Al Rahman F, Ibrahim Elnima E, Elhassan AM, Omar Hussein SE. Nanogel as a pharmaceutical carrier – Review article. Scholars Journal of Applied Medical Sciences. 2017;**5**(11F):4730-4736. DOI: 10.21276/sjams.2017.5.11.83

[48] Morishita M, Goto T, Nakamura K, Lowman AM, Takayama K, Peppas NA. Novel oral insulin delivery systems based on complexation polymer hydrogels: Single and multiple administration studies in type 1 and 2 diabetic rats. Journal of Controlled Release. 2006;**110**(3):587-594. DOI: 10.1016/j.jconrel.2005.10.029

[49] Alexander A, Ajazuddin K, Saraf JS, Saraf S. Polyethylene glycol (PEG)–poly (N-isopropylacrylamide) (PNIPAAm) based thermosensitive injectable hydrogels for biomedical applications. European Journal of Pharmaceutics and Biopharmaceutics. 2014;**88**(3):575-585. DOI: 10.1016/j.ejpb.2014.07.005

[50] Gulsen D, Chauhan A. Dispersion of microemulsion drops in HEMA hydrogel: a potential ophthalmic drug delivery vehicle. International Journal of Pharmaceutics. 2005;**292**(1-2):95-117. DOI: 10.1016/j.ijpharm.2004.11.033

[51] Jain S, Ancheria RK, Shrivastava S, Soni SL, Sharma M. An overview of Nanogel–novel drug delivery system. Asian Journal of Pharmaceutical Research and Development. 2019;**7**(2):47-55. DOI: 10.22270/ajprd.v7i2.482

[52] Morsi NM, Abdelbary GA, Ahmed MA. Silver sulfadiazine based cubosome hydrogels for topical treatment of burns: Development and in vitro/in vivo characterization. European Journal of Pharmaceutics and Biopharmaceutics. 2014;**86**(2):178-189. DOI: 10.1016/j.ejpb.2013.04.018

[53] Shimizu T, Kishida T, Hasegawa U, Ueda Y, Imanishi J, Yamagishi H, et al. Nanogel DDS enables sustained release of IL-12 for tumor immunotherapy. Biochemical and Biophysical Research Communications. 2008;**367**(2):330-335. DOI: 10.1016/j.bbrc.2007.12.112

[54] Wang Y, Cooke MJ, Morshead CM, Shoichet MS. Hydrogel delivery of erythropoietin to the brain for endogenous stem cell stimulation after stroke injury. Biomaterials. 2012;**33**(9):2681-2692. DOI: 10.1016/j.biomaterials.2011.12.031

[55] Wu W, Aiello M, Zhou T, Berliner A, Banerjee P, Zhou S. In-situ immobilization of quantum dots in polysaccharide-based nanogels for integration of optical pH-sensing, tumor cell imaging, and drug delivery. Biomaterials. 2010;**31**(11):3023-3031. DOI: 10.1016/j.biomaterials.2010.01.011

[56] Peng J, Qi T, Liao J, Chu B, Yang Q, Li W, et al. Controlled release of cisplatin from pH-thermal dual

responsive nanogels. Biomaterials. 2013;**34**(34):8726-8740. DOI: 10.1016/j.biomaterials.2013.07.092

[57] Vinogradov SV, Zeman AD, Batrakova EV, Kabanov AV. Polyplex Nanogel formulations for drug delivery of cytotoxic nucleoside analogs. Journal of Controlled Release. 2005;**107**(1):143-157. DOI: 10.1016/j.jconrel.2005.06.002

Chapter 6

Applications of Nanogel in Drug Delivery

Mansurat Oluwatoyin Shoge

Abstract

Drug delivery systems using nanogel are extremely essential. Chemical cross-linking is used to create it and the 3D polymer network of the nanogel has the capacity to encapsulate hydrophilic or hydrophobic therapies, such as proteins, compounds containing small molecules and ultrasmall nanoparticles. They were synthesized with a precise surface area and space due to their nanoscale structure, which also increased the stability of whatever medications they contained and increased the length of time they could circulate. Through the use of pH sensitivity, redox sensitivity, and temperature sensitivity, nanogels can achieve varied responsiveness. This is accomplished by designing specific chemical structures and employing various production methods. Consequently, the development of a multifunctional nanogel-based drug delivery system has increased the efficacy of illness therapies. As nanoscopic drug carriers, nanogels have drawn a lot of interest, especially for the site- or time-specific delivery of bioactive mediators. Nanogel preparations come in a variety of useful forms because to the wide variety of polymer systems and the straightforward adjustment of their physicochemical properties. Nanogels have exceptional levels of stability, drug loading potential, biologic consistency, strong permeation potential, and responsiveness to environmental cues. Nanogels have demonstrated great promise in a variety of sectors, including the delivery of genes, chemotherapeutic treatments, diagnosis, the targeting of particular organs, and many more. This review primarily focuses on various types of nanogels, preparation techniques, including techniques for loading drugs, various modes of biodegradation mechanisms, and primary mechanisms for drug release from nanogels.

Keywords: mediators, nanogels, polymer, cleavage, diagnosis

1. Introduction

A hydrogel is a three-dimensional network of a polymer that is cross-linked and hydrophilic, while a nanogel is a hydrogel where the particles are in the nanoscale size range. Nanogels can swell because they are hydrophilic, which explains why they can hold so much water. The ability to modify a nanogel's qualities by changing its composition is a crucial characteristic. Nanogels have a wide range of interesting uses, including gene delivery and medical imaging; in this case, the use of triggered drug administration will be taken into consideration. Due to their substantial

IntechOpen

swelling, nanogels may often entrap at least 30% weight of medicines through a variety of interactions. As the drug is absorbed, the gel breaks down and stable nanoparticles form, trapping the drug inside. A protective barrier surrounds the gel as the hydrophilic polymer chains at the surface become exposed, and this layer can be functionalized to target certain tissue or cells. Evidently, a nanogel's value as a triggered drug delivery system to a specific target depends on its compositional design and chemistry [1].

As nanoscopic drug carriers, nanogels have drawn a lot of interest, especially for the site- or time-specific delivery of bioactive mediators. Nanogel preparations come in a variety of useful forms because to the wide variety of polymer systems and the straightforward adjustment of their physicochemical properties. It has been demonstrated that nanogels have great potential in a variety of fields, including the delivery of genes, chemotherapy drugs, diagnosis, targeting of specific organs, and many others. This chapter will focus primarily on the various types of nanogels, methods of preparation, including methods of drug loading, and various modes of biological diffusion [2].

1.1 Types of nanogels

According to their cross-linking structure, nanogels can be divided into two main categories: chemically (covalent) cross-linked nanogels, which form cross-linking points through covalent bonds and physically cross-linked nanogels, which are cross-linked through non-covalent bonds like hydrogen bonds, electrostatic interactions, and hydrophobic interactions. They mostly resemble spheres. They can also be made with a cross-linked layer for structural integrity and either as a core-shell or core-shell-corona structure. Nanogels are extremely compatible with molecules that have a large loading capacity since they have a hydrophilic character.

1.1.1 Micelles

This type of nanogel is made from amphiphilic block or graft copolymers that have undergone supramolecular self-assembly in aqueous solutions. The centre of the micelle has adequate room to encapsulate pharmaceuticals and biomacromolecules. The effectiveness of the N-isopropylacrylamide-based micelle systems as drug delivery systems was examined [2]. It is made up of a core-shell morphological structure stabilized by hydrogen bonding, with a core hydrophobic block segment encircled by a hydrophilic polymer block shell (**Figure 1**).

1.1.2 PEG-PEI (poly(ethylene glycol)-polyethyleneimine)-based drug delivery

The initial nanogel to be introduced in 1999 was PEG-PEI. This is also an example of a nanogel. Since PEI (polyetheleneimine) had been demonstrated to be cytotoxic, PEG-PEI nanogels are often quite biocompatible. This is because of the PEG (polyethylene glycol). Due to PEG's proved non-toxic and non-immunogenic qualities, the material is made more soluble in water and less poisonous, making it more biocompatible than if PEI were used alone [2]. Numerous characteristics, including size, are crucial to understand because they influence a nanogel's potential applications and its efficacy as a drug release system. Due to the small size (20–220 nm) of PEG-PEI nanogel particles, penetration is probably improved.

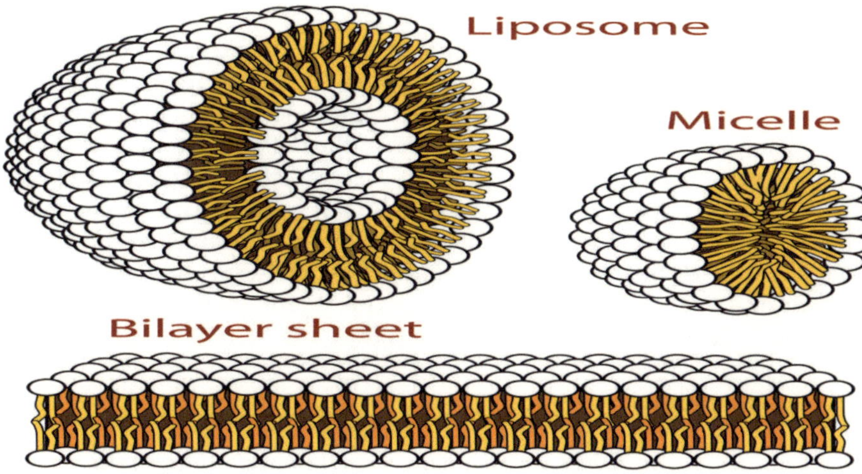

Figure 1.
Structure of micelle. Source: https://upload.wikimedia.org/wikipedia/commons/thumb/c/c6/Phospholipids_aqueous_solution_structures.svg/1200px-Phospholipids_aqueous_solution_structures.svg.png.

Since PEG-PEI nanogel particles are small (20–220 nm), it is expected that they will penetrate tissues and cells more readily, increasing the likelihood that the drug will affect its target.

Also, due to the charge of PEG-PEI, the amount of drug interaction between the nanogel and drug will affect how quickly the drug is released. Due to its high (positive) charge density, PEI is the ideal carrier for negatively charged pharmaceuticals and biological components [3]. Additionally, because pH can alter charge density, the release rate can be adjusted as the strength of the drug-nanogel connection changes. PEG-PEI nanogel applications in drug release systems are extremely promising and have the potential to have a significant impact on the biomedical industry.

It was reported that the encapsulation of AQ10 (6-(hydroxymethyl)-1,4-anthracenedione (AQ) analogue), a novel anti-cancer medication that targets pancreatic cancer [4]. When the medicine and nanogel were used together, cell proliferation was inhibited more efficiently, which is the desired result for cancer therapies, according to a comparison of taking the drug with or without the nanogel. The study came to the conclusion that the use of this nanogel as a drug delivery technology was successful because it made cells more receptive to the medication. PEG-PEI was employed in another investigation. In another study [3], SODN (antisense phosphorothioate oligonucleotides) distribution to multidrug-resistant (MDR) human oral epidermoid carcinoma cells was investigated using PEG-PEI nanogel. Overall, the study found that the medicine had a better chance of successfully accumulating in the cancer cells when PEG-PEI nanogel was present. This study is particularly notable because it introduces nanogels and establishes a baseline for all subsequent research on nanogels, whether it uses PEG-PEI or not [3].

1.2 Greater drug loading capacity of nanogels

The advantages of nanogels which have higher drug loading capacity are lesser carrier material, better control over the drug release and improved efficacy and safety [4]. The amount of drug loaded per unit weight of the nanoparticle is known as the loading capacity and it shows what proportion of the mass of the nanoparticle is made up of drug-encapsulated substance. By dividing the entire weight of the nanoparticles by the total amount of drug contained, one can get the loading capacity (LC%). The functional group that is present in the polymeric unit is what gives nanogels their greater drug loading capability. These functional groups play a significant role in the carrying and releasing abilities of drugs. Despite this, some functional groups have the capacity to combine a medication with an antibody for the purpose of targeting. The starting hydrogen bonding and van der Waal forces of attraction within the gel network are provided by these long functional groups of polymeric chains, as a result, reducing the carrying capacity for drugs [4].

1.3 Synthesis of nanogels

Physical covalent crosslinking or chemical covalent crosslinking was used to categorize nanogels. Some of the methods used to create nanogels include core-shell and hollow nanogel particles as well as controlled/living radical polymerization employing variable compositions, dimensions, and architectural designs. In order to introduce a high level of particular domain organization into nanogel and carry out a range of additional crosslinking operations, the core-shell self-assemblies, similar to polymer micelles, are used. Various distinctive features have recently been created in nanoscale fabrication techniques for fabricating precise nanogels. These characteristics include highly detailed control over size, shape, deformability, and surface chemistry [5] (**Figure 2**).

1.3.1 Nanogels for the delivery of small therapeutic molecules

Nanogels can quickly swell in an aqueous environment. The advanced design of the nanogel may be useful in adjusting the drug release rate, influencing carrier cell

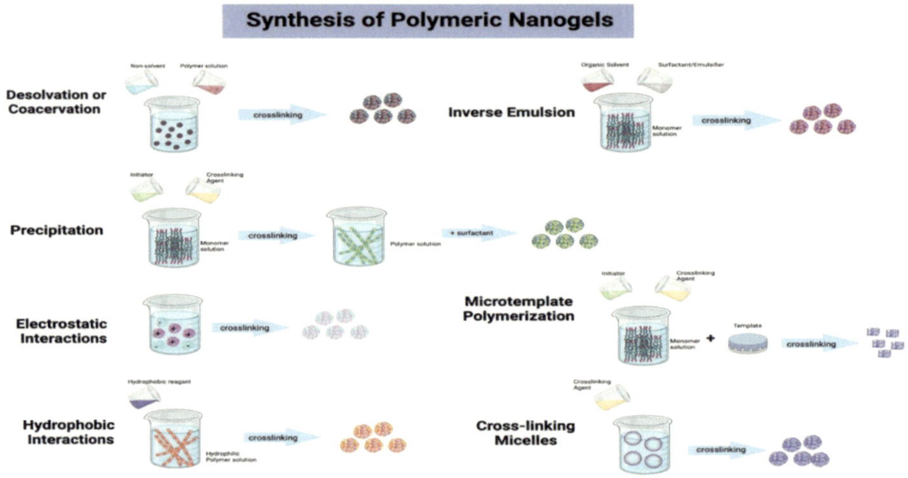

Figure 2.
Synthesis of polymeric nanogels. Source: https://www.researchgate.net/publication/289220718/figure/fig1/AS:388435031412737@1469621536175/Methods-of-nanogel-synthesis-the-polymer-precursor-method-and-the-emulsion-method.png.

interactions, and enhancing the medication's intended therapeutic effect. Weakly crosslinked polyelectrolyte nanogels' capacity to assimilate molecules with the opposite charge is one of their most notable characteristics. For instance, cationic crosslinked PEG-polyethyleneimine (PEG-PEI) nanogels were investigated for the immobilization of negatively charged physiologically active compounds such retinoic acid, Indomethacine, and valproic acid [5].

It was found that these drug-loaded nanogels reduced tumor development in breast cancer cells of an animal model by delivering the active triphosphates of therapeutic nucleoside analogs [4]. Recently, the same group investigated the advantages of cationic nanogel integrated active 5′-triphosphates of nucleoside reverse transcriptase inhibitors over free medicines in the antiviral therapy of HIV-1 infection in the central nervous system (CNS). The effectiveness of cationic nanogel integrated active 5′-triphosphates of nucleoside reverse transcriptase inhibitors over free medicines in the antiviral therapy of HIV-1 infection in the central nervous system (CNS) has recently been examined by the same group. Nanogels are macromolecular systems with the capacity to distribute their cargo at the desired location and to attain lengthy circulation half lifetimes in vivo. Nanogels are created with the idea of administration route in mind, and they get past any obstacles in their path to reach the circulation unharmed.

Nanogels are very expandable and can contain 30% weight. The polymer chain interacts with the biological molecules via electrostatic, van der Waal, or hydrophobic/covalent interactions. These loading capacities are unusually high and surpass those of liposomes and polymeric micelles. Following medication loading, the nanogels disintegrate into stable nanoparticles, trapping biohazard inside. Hydrophilic polymers (like PEG) can be introduced in a very thin structure to prevent aggregation. When the drug-nanogel complex breaks down, hydrophilic polymer chains are exposed, and they surround the nanogel in a protective layer. The versatility and control of polymer chemistry enable the creation and development of a wide variety of medication formulations as well as the incorporation of several therapeutic cargos inside the same nanogel carrier [4].

Dexamethasone is administered locally to avoid acute lung inflammation. The hybrid nanogel is crosslinked, composed of partially denaturated lysozyme cores and dextran shells, and is physically biocompatible. The pulmonary vasculature's target endothelial determinant was the target of antibodies coated on the nanogels.

1.3.2 In vivo activity

Due to their softness and deformability, splenic filtering is partially liquefied by nanogel. These are nanogel's primary characteristics. Due to their flexibility and deformability, erythrocytes, despite having a size range in microns, are able to readily pass through the splenic filtering bed, which has pores with a size of a few hundred nanometers.

There are various small compounds, peptides, antibodies, and other nanomedicines that can be delivered specifically to tissue or cells. Numerous tiny molecules, peptides, antibodies, or antibody fragments have been shown to be effective for the targeted administration of nanogel and other nanomedicine in tissue or cells. In comparison to nontargeted nanogel, ligand-mediated target alters the nanogel's biodistribution profile more. The strong expression of the receptor for the targeted nanogel can help to stop the over-deposition of the nanogel at the target site and so lessen the related adverse effects [5].

1.3.3 Using nanogel to distribute oligonucleotides

For the treatment and diagnostics of cancer and neurological diseases, therapeutic oligonucleotides were created for the targeted suppression of a certain mRNA (ribonucleic acid) sequence. These include the more recently identified micro RNAs (miRNAs), small interfering RNAs (siRNAs), and antisense oligodeoxynucleotides (ODNs). Realizing the ONs (oligonucleotides) delivery's full therapeutic potential is still greatly hampered by the inability to penetrate specific cells. ONs are hydrophilic molecules with a negative charge that are difficult to pass across cell membranes. They can activate the innate immune system and can be eliminated by endogenous nuclei. Therefore, ONs require a delivery truck in order to safely go to the point of action. One of the most well-known new types of nanomaterials for overcoming in vivo ONs delivery challenges is cationic nanogels [5].

1.3.4 Nanogels for the therapeutic delivery of proteins

Nanogels have been extensively researched for the transport of proteins and peptides in addition to their self-ability to encapsulate large quantities of biomacromolecules and prevent their disintegration. The molecular mass and hydrophobicity of the protein affect the formation of nanogels. The proteins' complexation, sizing, thermal denaturation, and ultimately aggregation shield the nanogel from enzymatic deterioration [6].

1.3.5 Delivery of nucleic acid

Small interfering RNAs, or siRNAs, have the potent capacity to selectively and effectively block gene expression. Small interfering RNA (siRNA) therapy is now a crucial component of treating diseases that are caused by genes. Due to fast enzymatic breakdown, it cannot penetrate the cell surface and has limited applications, such as modest transfection rates and brief half-lives.

In order to address these issues, siRNA can be loaded into liposomes, enhanced by the biomolecule cholesterol, or coupled to polymer nanoparticles while being processed with nucleic acids. A delivery mechanism based on nanogel recognized the siRNA treatment. Tetrahedral DNA-based nonviral vectors were introduced for siRNA assembly in nanogel, and they provide safety during delivery. This method might make it possible to effectively transfect cells both in vitro and in vivo while also stopping ribonuclease breakdown. In order to maximize productivity, this is the ideal environment for integrating different devices [7].

1.3.6 Nanogels for combined drug delivery

The nanogel structure is quickly modified to combine alternatives of different materials and, thus, provide advantages for combinatorial encapsulation of pharmaceuticals with varying chemical properties such small molecules, proteins, and nucleic acids. The development of liposomal nanogels composed of protein-encapsulating perishable polymers and drug-complexed cyclodextrins that may deliver soluble supermolecules (IL-2) and small hydrophobic molecules (TGF-). When given to mice, IL-2 and TGF- enhanced the lifespan of skin cancer tumors due to their synergistic effects [8].

When compared to single drug-loaded nanogels or free medication, binary drug combinations in nanogels demonstrated synergistic cytotoxicity against human ovarian A2780 cancer cells and exerted a stronger anticancer activity in cancer heterograft models used in-vivo [6]. The advantages of synchronic co-delivery of the platinum-taxane medication combination via a single carrier are increased by focusing on nanogels, which are overexpressed in the majority of ovarian tumors.

1.4 The use of nanogels in diagnostics and imaging

1.4.1 Nanogel for PET imaging

Researchers produced a polyacrylamide-based nanogel and crosslinked it with polydentate chelating ligands to create PET (Positron Emission Tomography) radiotracers by placing metal radionuclides on it [7]. For the purpose of enhancing the chelation stability of nanogels, a variety of crosslinkers including DTPA (diethylaminetriaminepolycarboxylic acid), DOTA (chelating agent), and 1, 4, 7-triazacyclononane-1,4,7-triacetic acid (NOTA) were produced. In contrast to the other 2 crosslinkers, NOTA-based nanogels preserved 64Cu the most consistently, according to experiments in mouse humor, with little to no trans-chelation. In comparison to DOTA-based nanogels, 64Cu-DOTA-crosslinked nanogels demonstrated substantial neoplasm accumulation as well as lesser signal in the liver and spleen. In several instances, the metastases' accumulation of 64Cu-DOTA was higher than that of the main connective tissue neoplasm. These findings suggest that metal-chelating crosslinked nanogels may be useful as PET agents for cancer detection and treatment monitoring. Such technologies have the advantage that the isotope can be quickly and easily integrated into premade nanogels prior to their clinical application [8].

1.4.2 Nanogel for optical imaging

Imaging is a crucial component of clinical protocols because it can offer morphological, structural, metabolic, functional, and molecular data for identifying and evaluating diseases. Imaging is a crucial component of clinical protocols because it can offer morphological, structural, metabolic, functional, and molecular data for identifying and evaluating diseases. The most widely used optical imaging techniques for nanogels are computed tomography (CT), positron emission tomography (PET), magnetic resonance imaging (MRI), and ultrasound.

High water content, structural flexibility, fluid-like transport, biocompatibility, and biodegradability are only a few of the characteristics of nanogels. By using metallic element ions to crosslink branching polyethyleneimines, gadolinium-assembled nanogels were created. To increase the duration of blood circulation, inverse microemulsion was used, followed by surface functionalization using synthetic resin glycol chains [8].

1.5 Nanogels' applications in drugs delivery

1.5.1 Antipyretic drug transdermal drug delivery

Aceclofenac's nanosized dispersion was created using emulsion-solvent diffusion techniques and added to a Carbopol 940. The formulation demonstrated ideal porosity characteristics, stability, and a sustained drug release.

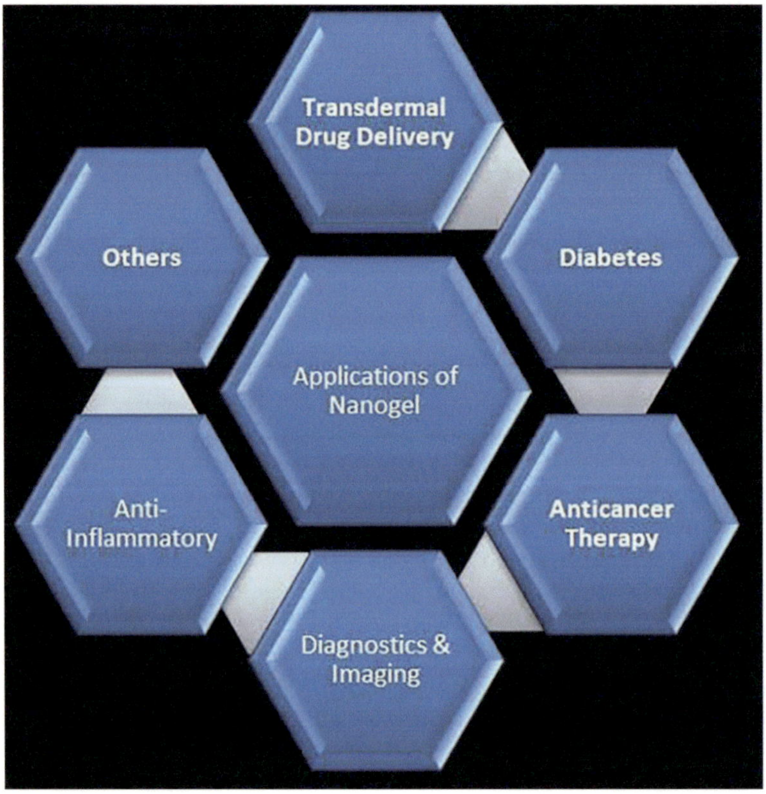

Figure 3.
Applications of nanogels. Source: https://www.futuremedicine.com/cms/10.2217/nnm.09.12/asset/images/medium/graphic47.gif.

It has been demonstrated that encapsulating magnetic nanoparticles like iron compound gives each mixture more stability and sensitivity than when these agents are delivered as unencapsulated substances [9]. Due to the cluster impact, nanogels enable the encapsulation of an excessive number of magnetic nanoparticles, which may result in the production of numerous, stronger native magnetic fields (**Figure 3**).

1.5.2 Vaginal medication delivery

1.5.2.1 Cancer treatment

In the recent years, researchers have looked at various drug delivery technologies, particularly nanoparticles, to overcome the drawbacks of conventional therapy agents such poor solubility, a limited therapeutic window, and toxicity [9]. An excellent benefit for the transport of anticancer drugs is expressed by nanogel, a type of unique nanoparticle. First off, the nanogel network's consistency offers a suitable cavern for storage when loading medications since it prevents early infection and acts as a barrier against environmental threats and deterioration. For instance, nanogel significantly preserved the potency of decitabine because of their improved stability and capacity to evade glycoside transporters. Second, the form of the carrier is changed by nanogels from a spherical to an oval shape that resembles a red blood cell. Due to the lengthening of blood circulation through microcapillaries, this characteristic is essential. The nanogels may lengthen under high shear rates, which would appear to make the blood less viscous. Additionally, nanogels will only serve as passive and active targets for tumor tissue [7].

To overcome the drawbacks of conventional therapy agents, such as poor solubility, a limited therapeutic window, and toxicity to conventional tissues, numerous drug delivery technologies, particularly nanoparticles, have been studied during the past few decades. An excellent benefit for the transport of anticancer drugs is expressed by the particular nanoparticle known as nanogel. The nanogel network's consistency firstly offers an ideal cavern for storage for loading medications in order to prevent premature unharnessing and serves as a barrier against environmental threats and degradation [8].

Additionally, the sensitivity and chargeability of nanogels are investigated as alternate targeting alternatives in cancer treatment. The effectiveness of the loaded drug will be significantly increased by pH-sensitive nanogels whether the drug acts extracellularly or intracellularly because, as was already mentioned, nanogels have a magnificent pH-responsive ability and growth may be a special tissue with a slightly acidic living thing microenvironment. Prepared PEG-chitosan nanogels for cancer treatment that are capable of being controlled by external cooling/heating as well as responding well to changes in pH scale setting [5]. Around the tumor extracellular pH scale (6.0–6.2), the 5-FU-loaded nanogels' surface charges ranged from virtually neutral to positive. The positive-charged nanogels will aid in cell learning, and eventually, the amplified acidity in endosomes and organelles will result in a significant release of 5-FU. By using synthetic polymers that can swell at acidic pH, temperatures above their transition temperatures, and reducing environments, it was possible to create a pH, thermal, and reaction potential triple-responsive expandable nanogel system (TRN) for the delivery of photo-sensitizers to target mitochondria and tumors. The nanogels rapidly enlarge from 108 nm to over 1200 nm (in diameter) during the course of 2 hours in a very reducing environment at blood heat. Additionally, TRN was functionalized with sigma-2 receptor targeting ligand, which is capable of efficiently focusing on head and neck cancer. The majority of the cancer cells were killed by TRN 12 hours after radiation, greatly enhancing the effectiveness of photodynamic therapy (PDT) in treating necrobiosis. In addition to efficiently delivering the hydrophobic photosensitizer, nanogel is also the optimum delivery mechanism for encasing the hydrophobic photosensitizer. The nanogel is the best delivery strategy for encasing the high number molecules (such as gold and platinum) that may enhance the biological impact of X-irradiation in addition to effectively delivering the hydrophobic photosensitizer. Numerous approaches to administering a curative dosage of radiation to neoplasia using nanogels have recently been studied. Developed PEGylated poly (2- [N, N, – diethylamino] ethyl methacrylate) nanogels to encapsulate highly stabilized and neoplasm specific gold nanoparticles [9].

A nanogel with a 106 nm diameter contained 15 gold nanoparticles (GNG). In addition to having a higher accumulating potency, the GNG in the nanogels had noticeably higher stability when compared to commercially available PEGylated GNG and Citrate-stabilized GNG. The gold-containing new PEGylated nanogel's results are conclusive. The findings demonstrate that a new PEGylated nanogel containing gold nanoparticles is a potentially effective nanomedicine for cancer photothermal treatment.

Nanoparticles will passively target cancer through EPR results, but often, their unfolding is limited to the outer region of the cancer and that they are unable to enter the deep cancer interstitial space.

1.5.2.2 Neurological conditions

A promising method for delivering oligonucleotides (ODN) to the brain may be nanogel. Systemic distribution of oligonucleotides (ODN) to the systema nervosum is necessary for the treatment of neurodegenerative illnesses. Blood-brain barrier (BBB) macromolecule injection results in poor translocation and rapid clearance from circulation. Nanogels that are certain or encapsulated with negatively charged ODN result in the creation of a stable liquid electrolyte dispersion complex with particle sizes under 100 nm that can be carried over the BBB with ease. Once siderophilin or insulin are added to the nanogel's surface, the transport effectiveness is further increased [10].

1.5.2.3 Anti-inflammatory activity

Nanogels can be used to administer anti-inflammatory drugs to people with skin issues. A skin penetrating nanogel system with a surface made of double-layered nanostructured particles and an emulsifying agent has been developed for efficient drug delivery in dermatitis. Nanogel was created using 3-acetyl-11-keto-boswellic acid (AKBA) and nanoparticles with AKBA loaded on them. The nanogels were organized using carbopol of the appropriate consistence. Nanoparticles with AKBA were used for treatment with ease [6].

The most potent pentacyclic triterpenic acid found in the gum of Boswellia serrata, 3-acetyl-11-keto-boswellic acid (AKBA), exhibits anti-inflammatory properties. In AKBA, there is a lot of higher medicine activities. These studies shown that, in cases of skin irritation, medications can be administered successfully through the skin. Non-steroidal anti-inflammatory drugs (NSAIDs) of the diclofenac class are frequently used to treat arthritis. Diclofenac sodium self-assembling gel hydrogels produced better anti-inflammatory effects. Additionally, it has been discovered that nanogels containing cinnamon oil and cinnamaldehyde have strong antibacterial properties [10].

1.5.2.4 Additional uses

A nanogel made of polyvinyl pyrrolidone and poly(acrylic acid), or PVP/PAAc, is sensitive to the concentration of hydrogen ions. The pre-ocular retention and ocular penetration abilities of the curcumin are enhanced by using film-ultrasonic methods and thermosensitive gelling agents in curcumin-loaded cationic nanostructured supermolecule carriers (CNLC). Muscone has a colloidal gel drug loading capability, and physical studies indicated that the phase change temperature was 34°C. Based on initial in-vivo investigations, insulin-loaded nanogels seem to be 51% more successful than free insulin at preserving blood glucose levels in diabetic rats [7].

1.5.2.5 Immune disorder

In a study, a novel nanogel drug delivery system for the immunosuppressive medication mycophenolic acid (MPA) was developed and tested [6]. The findings of this study showed that nanogel-based native medication delivery is substantially more effective for treating lupus erythromatous because it specifically targets antigen-presenting cells. The innovative medicine delivery system will prolong the patient's life and postpone the onset of renal failure.

1.5.2.6 Stopping bleeding

Major wound bleeding has been controlled using a protein molecule found in the solution that was utilized to create a nanogel. The proteins have a way for self-assembling into a biodegradable gel at the nanoscale. For instance, micronized sacchachitin promotes wound healing [8].

1.5.2.7 Nasal drug administration

Technologies for delivering drugs via nanogels have a great deal of potential for overcoming some of the problems with medication delivery. Because the nasal mucosa absorbs nanogels so quickly, they could be used to transport and distribute medications through the mucosa. An inventive method to slow the spread of disease is nasal immunization with nanogels [7]. The use of nanogels for vaccine distribution via the nasal route may be a novel way to control the spread of a disease. A composite polymeric network of nanoparticulates (NPs, microcapsules, NEs, etc.) is what gives nanogels their high viscosity.

1.6 Advantages and disadvantages of nanogels

The main benefit of nanogels is their small size since they offer a lot of specific surface area, which increases the possibility of interaction between the components. This is advantageous since it increases the likelihood that the medicine will reach the intended target and produce the desired effect. Nanogels can also be altered to be used for a variety of purposes because of their tunability. Nanogels have a wide range of applications, as shown by one study that functionalized the surface of PEG-PEI by adding an amine group, enabling the nanogel to be utilized for treating spinal cord injuries [7].

There aren't many drawbacks to using nanogels for drug delivery. One illustration is that because it is diffusion-based, the release rate may be too rapid. Nevertheless, numerous studies have concentrated on resolving this issue by ensuring that there is a contact between the medicine and gel, delaying the release.

One major advantage of nanogels is that the research is always changing; as in the case of the example presented, as soon as a possible problem is detected, a remedy is quickly proposed and studied.

2. Conclusion

Nanogels are reassuringly modern drug delivery systems that effectively balance the drawbacks of both traditional and contemporary therapies, such as low stability and nonspecific effects. Biological ligands such proteins, peptides, and polysaccharides were produced with modifications to nanogel for optimal contact with target cells. Because of their high incorporation capacity, swelling ability, and the quantity of biological molecules interacting with the electrostatic, van der Waal interaction that occurs with polymer chains, leading to a stable nanoparticle and allowing for easy drug entrapment, nanogels are used as therapeutic drug carriers. The use of nanogels for tumor targeting appears to be quite effective. In order to promote highly selective absorption into specific cancer cells and reduce their uptake in normal cells, future objectives of this review will thus be to improve the design of nanogel with specific targeting residue.

Therefore, it can be said that because nanogel particles are flexible, they make an excellent platform for triggered drug administration. This enables them to have qualities that are advantageous for a variety of targeted release systems. As previously said, current research confirms the effectiveness of PEG-PEI nanogels and other nanogels in triggered drug delivery [8]. Although nanogel research has advanced significantly over time, more study is still required before clinical application.

Author details

Mansurat Oluwatoyin Shoge
Air Force Institute of Technology (AFIT), Kaduna, Nigeria

*Address all correspondence to: moshachemist@yahoo.com;
m.shoge@afit.edu.ng

References

[1] Vinogradov S, Batrakova E, Kabanov A. Poly(ethylene glycol)–Polyethyleneimine NanoGel™ particles: Novel drug delivery systems for antisense oligonucleotides. Colloids and Surfaces B: Biointerfaces. 1999;**16**(1-4):291-304. DOI: 10.1016/S0927-7765(99)00080-6

[2] Mauri E, Chincarini G, Rigamonti R, Magagnin L, Sacchetti A, Rossi F. Modulation of electrostatic interactions to improve controlled drug delivery from nanogels. Materials Science and Engineering: C. 2017;**72**:308-315. DOI: 10.1016/j.msec.2016.11.081

[3] Pinelli F, Pizzetti F, Ortolà Ó, Marchetti A, Rossetti A, Sacchetti A, et al. Influence of the core formulation on features and drug delivery ability of carbamate-based nanogels. International Journal of Molecular Sciences. 2020;**21**(18):6621. DOI: 10.3390/ijms21186621

[4] Soni K, Desale S, Bronich T. Nanogels: An overview of properties, biomedical applications and obstacles to clinical translation. Journal of Controlled Release. 2016;**240**:109-126. DOI: 10.1016/j.jconrel.2015.11.009

[5] Pinelli F, Pizzetti F, Rossetti A, Posel Z, Masi M, Sacchetti A, et al. Effect of surface decoration on properties and drug release ability of nanogels. Colloids and Surfaces A: Physicochemical and Engineering Aspects. 2021;**614**:126164. DOI: 10.1016/j.colsurfa.2021.126164

[6] Ganta C, Shi A, Battina S, Pyle M, Rana S, Hua D, et al. Combination of nanogel polyethylene glycol-polyethylenimine and 6(hydroxymethyl)-1,4-anthracenedione as an anticancer nanomedicine. Journal of Nanoscience and Nanotechnology. 2008;**8**(5):2334-2340. DOI: 10.1166/jnn.2008.294

[7] Vismara I, Papa S, Veneruso V, Mauri E, Mariani A, De Paola M, et al. Selective modulation of A1 astrocytes by drug-loaded nano-structured gel in spinal cord injury. ACS Nano. 2019;**14**(1):360-371. DOI: 10.1021/acsnano.9b05579

[8] Cuggino C, Blanco E, Gugliotta L, Alvarez I, Cecilia I, Calderon M. Crossing biological barriers with nanogels to improve drug delivery performance. Journal of Controlled Release. 2019;**307**:221-246. DOI: 10.1016/j.jconrel.2019.06.005

[9] Suhail M, Rosenholm J, Minhas M, Badshah S, Naeem A, Khan K, et al. Nanogels as drug-delivery systems: A comprehensive overview. Therapeutic Delivery. 2019;**10**(11):697-717. DOI: 10.4155/tde-2019-0010

[10] Li C, Obireddy S, Lai W. Preparation and use of nanogels as carriers of drugs. Drug Delivery. 2021;**28**(1):1594-1602. DOI: 10.1080/10717544.2021.1955042

Chapter 7

The Potentials and Challenges of Hydrogels for Ocular Therapy

Chukwuebuka Umeyor, Emmanuel Uronnachi, Abhit Nayak, Tochukwu Okeke, Purav Shah and Vandana Patravale

Abstract

The major objective of any pharmacotherapeutic activity is to achieve an effective concentration of drug at a particular site of action for a sufficient period of time to produce a desired response or effect. The eye is a very important organ of the body because of its use in vision as well as its easy accessibility. Though solution-type drug delivery to the eye records high patient adherence but it is limited by poor ocular bioavailability due to certain pre-corneal physiological and anatomical obstacles. Hydrogels are important self-assembly nanoformulations that serve as alternatives to solution-type eye preparations with good potential to produce enhanced local absorption and bioavailability in the treatment of eye disorders that may be vision-threatening. This chapter will present an overview of the eye anatomy, ocular barriers, hydrogels and their classes, applications in ocular diseases, and future prospects of hydrogels in ophthalmic therapeutics.

Keywords: eye, ocular delivery, ocular barrier, hydrogel, nanomedicine, ocular diseases

1. Introduction

The eye is an important organ of the body due to its prominent role as the centre of vision which helps to keep the body safe from physical injury and harm. Sometimes, the visual functions of this vital organ is compromised by a variety of pathologic conditions and injuries such as bacterial or fungal infections, vascular diseases of the retina, inflammatory diseases, tumors, glaucomatous neuropathies, cataract, dry eye, and macular degeneration, which may lead to impaired or total loss of vision [1]. Various ophthalmic interventions including solutions, drops, emulsions, gels, lubricants, ointments, films, and implants incorporating small drug molecules and peptides have been designed, developed, and applied for the clinical management of these ocular pathologies [2]. The pros and cons of these treatment strategies have been reviewed and discussed in several reports. The major concern raised in these reports is the negative impact of the lacrimal drainage and tear dilution system of the eye on drug delivery because they facilitate rapid clearance of applied therapeutics resulting in decreased pre-corneal retention time, low bioavailability, and poor therapeutic outcome [3]. To advance efficient ocular drug delivery for alleviation or elimination of these conditions, researchers have focused their interests in the fabrication

IntechOpen

of bioinspired delivery systems using materials that have the ability to maintain prolonged contact with the compromised visual architecture of the eye. This would ensure that the long term treatment goals of topical drug administration to the eye are realized. The topical route is one of the easiest and commonest routes of drug administration. However, topical ocular delivery mostly favors the anterior segment of the eye with little or no impact felt in the posterior eye due to poor accessibility and the complexities of the physiological and anatomical structures of the eye. In addition, increasing the dose by intravitreal or intravenous administration will increase toxicity risks due to the possibility of drug uptake in the conjuctiva and systemic dose dumping through the nasolacrimal pathway [4]. Therefore, the ultimate objective of innovative biomimetic delivery systems like liposomes, nanotubes, nanoparticles, micelles, contact devices, hydrogels, and dendrimers, is to improve permeation, surface contact, pre-corneal retention time, and deposition of encapsulated payloads to both the anterior and posterior segments of the eye.

Hydrogel is a hydrophilic drug delivery platform fabricated into a three-dimensional array using swellable polymers which possess excellent water absorption and retention capacity. The concept of hydrogel was first hinted in the late nineteenth century while an attempt to formerly describe it was made in the early twentieth century using a ternary system comprising water, alcohol, and gelatin [5]. With recent advances in the chemistry of polymer synthesis, smart and stimuli-reactive hydrogels have been synthesized incorporating drugs, genes, biologics, and other small molecules for targeted ophthalmic therapy by enhanced co-administration of drugs, prolonged contact and residence times at the target site, and controlled drug release in the ocular tissue [6]. Often, hydrogels have been synthesized to serve as the core carrier system for other nanoparticulate systems like solid lipid nanoparticles, lipid nanoemulsions, liposomes, or nanostructured lipid carriers for therapeutic purposes. Their properties could be tuned by covalent or physical crosslinking of ligands and molecules to improve their surface, mechanical, retention, safety, stimuli-responsive, and targeting profiles [7]. This chapter will present an overview of the eye anatomy, ocular barriers to drug delivery, discuss different classes of hydrogels and their fabrication techniques, explore mechanisms of drug entrapment in hydrogels and their biocompatibility with the eye, applications of hydrogels in ocular diseases, challenges in hydrogels translation for ocular application, and future prospects of hydrogels in ophthalmic therapeutics.

2. Anatomic and physiologic features of the eye

The eye is structured into two distinct parts: the anterior and posterior parts of the eye. The anterior eye comprises the cornea, iris, conjuctiva, crystalline lens, aqueous humor, and ciliary body, whereas the posterior portion of the eye is made up of the neural retina, choroid, retinal pigment epithelium, vitreous humor, and the sclera. The accessory structures of the eye include the eyelids, eyelashes, eyebrows, the lacrimal (tear-producing) apparatus, and extraocular muscles. The eyeball is rooted into the cranium and its movement is regulated by the extraocular muscles (**Figure 1**) [8, 9].

2.1 Anterior eye

The cornea is the most sensitive, highly specialized, and innervated tissue of the anterior eye which measures 10.5 by 11.5 mm in a normal healthy adult. The cornea

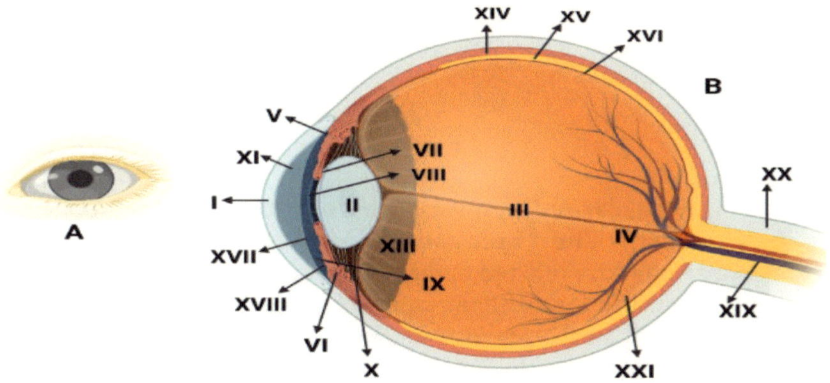

Figure 1.
(A) Frontal view of the human eye. (B) Illustration of the anatomical structures of the eye: I—cornea; II—lens; III—vitreous body; IV—optic disc; V—conjunctival fornix; VI—sclera; VII—iris; VIII—anterior chamber; IX—iridocorneal angle; X—ciliary body; XI—meibomian glands; XII—posterior chamber; XIII—suspensory ligament; XIV—choroid; XV—retinal pigmented epithelium; XVI—retina; XVII—palpebral conjuctiva; XVIII—bulbar conjunctiva; XIX—central artery and vein of the retina; XX—optic nerve; XXI—fovea.

is stratified into the epithelium, lamellar stroma, Bowman's layer, Desçemet's membrane, and endothelium. The stroma supports the mechanical structure of the cornea. The epithelium serves as a barrier to the permeation of toxins into the ocular intracellular chamber. The Bowman's layer or membrane lack cells, does not regenerate and forms scars when damaged. The Desçemet's membrane is usually produced by endothelial cells *in utero* during which it is banded or after birth when it is amorphous or unbanded. The endothelium maintains corneal transparency, dehydration, and clarity necessary for enhanced vision [10]. The conjuctiva has three components with distinct functions—palpebral, forniceal, and bulbar conjunctiva. Forniceal forms a barrier to the exposed eye, and the palpebral conjunctiva holds the eyelids in position. The conjunctiva is responsible for ocular immunity (provided through the conjunctiva-associated lymphoid tissue), motion due to its elastic nature, protection, and lubrication [8]. Conjuctiva epithelium contains Manz glands, mucous glands, Henle crypts, and goblet cells which produce mucin, and electrolytic fluids components of tear film. The cells are present in the palpebral, forniceal, and bulbar conjunctiva [11]. The iris is a brightly colored layer which regulates the amount of light that enters the eye. It serves as the root of formation of the ciliary body, and it makes a spherical impression on the lens, referred to as the pupil, which adjusts its size according to available light, controls light penetration through the retina, and maintains visual acuity [12]. The ciliary body is very closely localized with the iris. The capillaries of the ciliary body facilitates exchange of impulses and sensory information between the anterior and posterior eye. The ciliary body is involved in aqueous humor secretion, and assists the lens to adjust its focus on objects [13]. The aqueous humor is secreted from plasma in the ciliary body at the rate of about 2.5 μL/min through active production, ultrafiltration, and diffusion which ensures effective distribution and draining. The average period of turnover for aqueous humor is about 100 minutes. It is a rich source of glucose, ascorbic acid, globulins, albumin, oxygen, and other essential nutrients for the cornea and crystalline lens, and also evacuates debris, wastes materials, and toxins from these non-vascularized tissues. It is also responsible for the maintenance of intraocular pressure (IOP) and shape of the eye [8]. Crystalline lens is a biconvex, non-innervated ocular tissue that is held in position by the zonula fibers of the ciliary body. The lens controls the transfer of waste products and metabolic materials through diffusion. It is compartmentalized into the nucleus, epithelium, capsule, and cortex [14].

2.2 Posterior eye

The choroid is a vascularized layer located between the sclera and retina which contains mucinated extracellular fluid and melanocytes. The choroid receives a high amount of vascular supply compared to the brain and other parts of the eye. The high blood supply is responsible for excellent supply of nutrients and oxygen to the retina, the removal of metabolic wastes and maintenance of IOP [13]. The normal human eye contains about 3.5 million retinal pigment epithelial cells which converge into a tight junction called zonulae occludentes. These cells proliferate in disease conditions because they produce large amount of growth factors including ciliary neurotropic, vascular endothelial, and platelet-derived growth factors. The cells secrete catalase, melanin pigment, superoxide dismutase, and immunomodulatory enzymes which protect the eye against oxidative damages. It supports the photoreceptor cells (amacrine, rod, cone, bipolar, horizontal, ganglion, interplexiform, and glia cells) to maintain effective vision cycle and control of chemical components of the retina [15]. The sclera (white eye) is a mesh-like continuation of the corneal tissue which contains mucopolysaccharides, fibroblasts, and collagen fibers from the central nervous system. It gives shape to the eyeball, makes it more rigid, protects its inner parts, and serves as a site of attachment for the extrinsic eye muscles [13, 16]. The neural retina is made up of millions of photoreceptor cells mainly rod and cone cells, and facilitates the exchange of sensory information between the brain and the ocular exterior. The rod and cone cells trap visual nerve information which is relayed to the brain through bipolar cells, and support differentiation of colored objects in low and bright light. The vitreous humor (or vitreous body) is rich with water and has negligible amounts of hyaluronic acid and ions, and collagen fibers. The quality of the vitreous body deteriorates with age [13, 15].

3. Ocular drug delivery barriers

3.1 Corneal barrier

Generally, topical drug transport into the cornea is influenced by the pH and degree of ionization of the drug molecules. The cornea presents sufficient permeation obstacle to topical dosage forms due to its physiomechanical features especially its content of epithelial cells [17]. The epithelial cells hinder drug absorption due to their tight adherence to ocular proteins like occludens (Zonula occludens-1 and Zonula occludens-2), and cingulin, forming tight junctions which decrease paracellular transportation of hydrophilic or ionic drugs into the cornea while transcellular-mediated diffusion propels the permeation of hydrophobic drugs. The corneal stroma acts as a barrier to the deep absorption of lipophilic and modest hydrophilic drugs by trapping the molecules as a reservoir. The stroma forms a barrier with endothelium but this obstacle possesses leaky structures which permits the passage of macromolecules. The barrier function of epithelial cells or tight junctions of the cornea is lost in pathologic conditions such as vernal keratoconjuctivitis (**Figure 2**) [18].

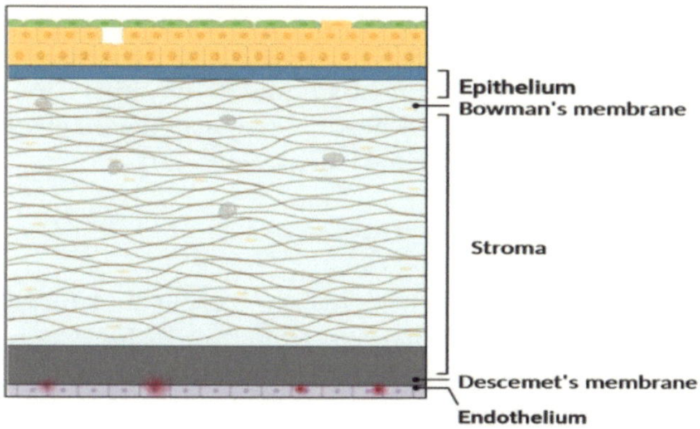

Figure 2.
Ocular drug delivery barriers of the cornea (reproduced from [18] under the terms and conditions of the creative commons attribution (CC BY) license (https://creativecommons.org/licenses/by/4.0/).

3.2 Blood aqueous barrier

This barrier comprises the ciliary epithelium, blood vessels of the ciliary body, and epithelial cells of the iris. These ocular factors form junctions which limit or regulate the access of non-targeted drug molecules to the deep tissues of the eye. Through this mechanism, the tight junctions assist the eye to maintain its homeostatic chemical components and visual clarity [19].

3.3 Conjunctival barrier

Drug uptake through the conjunctiva following topical application is implemented through the transcellular, endocytic, paracellular, and active routes of absorption. However, the transepithelial electrical resistance acts a barrier to drug uptake across the conjunctiva. The paracellular pathway is decorated with tight junctions which act as obstacle to drug absorption. Drug absorption in the conjunctival tissue is also negatively impacted by secretory cells in the tissue. Tear film acts as a barrier to the concentration and bioavailability of drug molecules absorbed in the conjunctiva. Generally, molecular weight and aqueous solubility are two vital determinants of drug uptake in the conjunctiva. Water-soluble drugs with molecular weight < 20 kDa are more likely to be absorbed than molecules with higher molecular weight. Further, conjunctiva is a centre of excellence for the absorption of proteins and peptides such as insulin. Its uptake is facilitated by paracellular mechanism of drug transport due to the leaky microstructure and large surface area of the conjunctiva [19].

3.4 Barrier due to efflux proteins

Efflux proteins located in the lateral or apical region of the cell membrane act as barriers to drug absorption. Two prominent examples include a member of the ABC (ATP-binding cassette) protein known as multi-drug resistant protein, which acts as a barrier to the transport of conjugated and organic molecules, and p-glycoprotein which hinders the absorption of amphipathic drugs. P-glycoprotein 1 has an approximate molar weight of 170 kDa found in polarized cells and it decreases drug permeation in multi-drug resistant cells. It is secreted by ciliary-iris muscles, ciliary epithelium, epithelial cells of the conjunctiva, capillary endothelial cells of the retina, and cornea. Multi-drug resistant protein is an efflux protein with an approximate molar weight of 190 kDa that is prominently found in the kidney, liver, and intestines. Its transporter efficiency is improved when it is conjugated with glucuronides, bile salts, glutathione, sulphate, and cysteinyl leukotrienes [20].

3.5 Nasolacrimal production and drainage

Tear film is made up of mineral salts, lysozymes, water, and antibodies. It aids the eye to produce clear vision, wash off irritants from the eye, lubricates the eye, maintain a strong immune response, and healthy ocular epithelial cells [21]. Topical drops which are instilled to the eye are docked in the cul-de-sac of the conjunctiva and most of them are eliminated into the nasolacrimal ducts through precorneal tear drainage system due to prior adherence of drug to tear proteins, tear dilution, and accelerated drug clearance process. The lacrimal system is the pathway through which tear produced in the eye flow into the nasal cavity, and the system is made up of lacrimal sac, canaliculi, nasolacrimal duct, and puncta. Tear drainage due to reflex blinking (about 6 blinks/min) restricts drug permeation and acts as a barrier to overall drug available for absorption and therapeutic effect. Drug drainage from the precorneal ducts occur within 2 minutes resulting in decreased ocular contact time and low bioavailability of absorbed drug [8].

3.6 Aqueous humor barrier

The flow of aqueous humor in opposite direction to that of drugs instilled topically to the ocular tissue acts as an obstacle to its absolute absorption. In addition, there is increased drainage of hydrophilic drugs into the Schlemm canal through the trabecular mesh. This barrier function which results in the bioavailability of sub-therapeutic concentrations of instilled drugs could be limited by pathologic conditions like pathogenic inflammation of the ocular tissue due to decreased drainage of topically administered drugs. However, this condition restricts the permeation and absorption of drugs administered to the eye through intravitreal injection as already demonstrated in a report using fluorescence-labeled albumin [22].

3.7 Scleral barrier

Sclera barrier is one of the prominent static barriers of the posterior eye. Due to its protective functions, drug transport across the sclera depends on the physicochemical profile, water solubility, molecular weight, and surface charge of the drug molecules. Thus, poor aqueous solubility and increased molecular weight of drug molecules results in limited delivery across the scleral pores and low absorption. Also, drug molecules with opposite surface charge to the negative charge on the scleral pores experience decreased permeation and are entrapped in the pores [23].

3.8 Blood retinal barrier

The blood retinal barrier selectively regulates the transport of drug molecules following periocular or systemic administration to the retina. The barrier has two layers—the outer layer comprising the retinal pigment epithelium (or epithelial cells), and the inner layer which houses the endothelial cells. These layers possess tight junctions (supported by Müller cells and astrocytes) which selectively control the transverse of biomolecules present in the blood into and out of the vitreous environment mainly through a passive transcellular transport system. In this system, permeability of drug molecules is inversely related to the molecular weight of drug as reported in a study using bovine retina. The efficiency of the retinal transport mechanism is maintained by transporter moieties such as organic anion polypeptides, and organic cation and anion transporters, and these transporters play prominent roles in the ocular distribution of proteins, peptides, biologics, and hormones in the retinal milieu [24].

3.9 Hydrogels for ocular therapeutics

Conventional methods of ocular drug delivery which include the use of drops and ointments have faced numerous challenges ranging from poor ocular bioavailability to rapid clearance from administration site. These hindrances are the result of unique physiological features inherent to the eye anatomy. Hydrogels in ocular therapeutics tend to provide cogent solutions to the problems encountered by conventional ocular drug delivery systems. Their ability to attain high biocompatibility, drug loading and bioavailability in the eye make them an ideal choice for ocular drug delivery [25]. Hydrogels are made up of three dimensional polymeric chain networks which have the ability to swell in aqueous solvents and hold other substances such as drugs, proteins, and other molecules. They are so versatile and can be specifically designed to exploit environmental physiological conditions to illicit a predetermined characteristic behavior at the administration site. Hydrogels which possess such ability are called "stimuli-sensitive" hydrogels because they detect physical, chemical or biological changes in physiological conditions and react by altering swelling capacity. This allows for both sustained and controlled release as well as optimal ocular bioavailability [26]. Changes in pH and ion concentrations of the physiological environment make up the chemical factors that influence hydrogel behavior. Other factors such as light pressure, sound and temperature are often referred to as physical stimuli. Hydrogels can also be designed as *in situ* gelling systems where the hydrogel is liquid before administration and turns to a gel after administration, thus undergoing a sol-gel transition. This involves stimuli sensitivity; therefore, gelation post-administration can occur as a result of changes in pH, temperature, or ion concentration. Fabrication of hydrogels with hydrophobic molecules can be a bit challenging because of the incompatibility existing between the molecules and aqueous contents of hydrogels. However, methods have been developed to integrate such drugs into hydrogels for ocular delivery. Hydrogels for ocular drug delivery are already in use, one of which is the soft disposable contact lens. Soft contact lenses (SCLs) are hydrogels that can increase corneal residence time of ophthalmic drugs thus increasing bioavailability [27]. These lenses are usually made from conventional polymers such as poly-(2-hydroxyethylmethacrylate) (pHEMA), silicone, gelatin, chitosan, hyaluronic acid, alginate, methylcellulose and collagen. Silicone SCLs can be employed for extended use due to increased oxygen permeability. Unlike hydrophobic drugs, hydrophilic drugs can be easily incorporated into SCLs by soaking the already formed SCLs in a solution of such drug. The use of organic solvents to dissolve hydrophobic drugs has been mentioned by some researchers as an effective method to incorporate hydrophobic drugs in SCLs. Latanoprost was loaded in silicone SCLs by soaking the lenses in a solution of latanoprost in n-propanol [28]. Other methods which include the use of colloidal systems such as micelles and microemulsions have also been employed to counteract the solubility challenges of hydrophobic drugs. This, however, may reduce the average light transmission of the hydrogel depending on the globule sizes of the colloidal system [29]. Bimatoprost was loaded in pHEMA SCLs by soaking in drug loaded microemulsion. *In vitro* release profile of this SCLs showed an extended drug release of up to 48–96 hours which is greater than the release observed in SCLs without microemulsion [30]. Other methods employed to improve drug loading and release form SCLs include molecular imprinting with hydrogels and supercritical solvent impregnation [27]. Molecular imprinting involves the introduction of artificial receptors for the target drug to be loaded on the hydrogels. Hydrophobic drugs like bimatoprost and prednisolone loaded in SCLs with this method have shown increased drug loading capacity and release durations [31]. Supercritical solvent impregnation relies on the unique properties of supercritical fluids to improve the perfusion of the dissolved hydrophobic drug into SCLs. Acetazolamide, an anti-glaucoma drug was incorporated in SCLs using this technique and a drug release duration of 450 minutes was reported [32]. Polymers employed in the fabrication of hydrogels for ocular delivery are biodegradable and do not cause inflammations during use. Chitosan, a cationic polysaccharide derived from chitin possesses mucoadhesive properties as a result of the interaction between its positive charge and the negatively charged mucin [33]. This enables drug permeation and controlled release of drug into the ocular tissues. However, due to its poor alkaline solubility, it is usually modified through PEGylation or carboxymethylation [34]. It can also be combined with other natural or synthetic polymers to optimize ocular drug permeability and release. Hyaluronic acid is another polymer which has shown immense potential in ocular drug delivery because it is a component of the vitreous humor of the eye [35]. It is highly biocompatible and non-immunogenic due to its natural occurrence in the body. However, it cannot form a gel on its own; so for it to be used for hydrogels, it must be either chemically modified or combined with a gelling agent. Gelatin polymeric platforms have also been explored in ocular drug delivery. Timolol hydrogels fabricated with gelatin and chitosan in a study by Song et al. [32] showed greater reduction in intraocular pressure and longer duration of action than the commercially available product. Alginate polymers undergo ion-responsive gelation. They are also biocompatible, biodegradable and not toxic to human tissues. Methylcellulose is also vastly used majorly as a viscosity enhancer in ocular delivery systems and more importantly, it is employed in *in situ* gelling hydrogel systems because of its heat-induced reversible sol-gel transition property [36].

4. Fabrication of hydrogels

Hydrogels are formulated by the crosslinking of polymers mainly initiated by physical or chemical means. The different methods for fabrication of hydrogels are discussed below.

4.1 Physical crosslinking

This involves the fabrication of hydrogels through the formation of non-covalent interactions along the polymer chains. These interactions can be as a result of electrostatic interactions, hydrogen bonding, and hydrophobic forces. These hydrogels are called physical gels because their formation was initiated by physical factors; thus, are reversible. They are highly sensitive to water and temperature [37]. They have a short lifespan in the physiological media and are therefore used for short term delivery of drugs. They do not contain toxic chemicals and are quite safe to use. Polymers are mixed together in appropriate conditions and interactions between polymer molecules are initiated. Hydrogels based on polysaccharide polymers such as chitosan are prepared using this method. Chitosan interacts with anionic molecules such as phosphates and sulfates to form physical hydrogels. These anions interact with chitosan through a protonated amino group. The concentration of chitosan as well as its degree of interactions with available anion molecules determine the properties of the hydrogel formed. Other naturally occurring polymers such as alginate and pectin also undergo ionic interactions to form physical hydrogels. Alginate can interact with Ca^{2+} ions while carrageenan interacts with K^{+} ions to form physical hydrogels. Melded non-ionic polymers undergoing freeze-thaw cycles can also produce physical hydrogels. The hydrogel is formed possibly as a result of crystallization reaction occurring via hydrogen bonding within the molecules. Polyvinyl alcohol (PVA) aqueous solution transformed into hydrogel after several freeze-thaw cycles and the hydrogel properties were determined by the concentration of the PVA and number of freeze-thaw cycles [38].

4.2 Chemical crosslinking

Unlike physically crosslinked hydrogels, chemically cross linked hydrogels can be easily manipulated to exhibit a pre-determined behavior. They are irreversible because they are formed by covalent interactions between polymer molecules. Physical properties of the chemical hydrogels such as mechanical strength and swelling capacity can be altered to achieve desired results [38]. There are different methods employed in the preparation of hydrogels through chemical cross-linking and they are as follows:

4.2.1 Use of chemical crosslinker

This involves covalent bonds mediated by hydrophilic groups. A schiff base is usually formed indicative of the covalent complexation between polymer chains. These hydrophilic groups usually include the hydroxyl, carboxylic, and amino groups [39].

4.2.2 Crosslinking by addition reaction

This is the reaction of hydrophilic polymers with higher functional crosslinkers to form larger polymer with longer chain length. These crosslinkers, some of which include 1,6-hexamethylenediisocyanate and 1,6-hexanedibromide have been used to crosslink polysaccharides producing hydrogels [40].

4.2.3 Crosslinking with aldehydes

This involves the use of aldehydes to crosslink hydrophilic polymers containing hydroxyl groups. This usually occurs at high temperature and acidic pH with methanol added to quench the reaction. This is observed in the crosslinking of polyvinyl alcohol with glutaraldehyde [41].

4.2.4 Crosslinking by condensation reaction

Reactions between amino or hydroxyl group with derivatives of the carboxylic group have been reported to produce hydrogels. Gelatin hydrogel was prepared using 1-ethyl-3-(3-dimethylaminopropyl)carbodiimide, a good cross-linking reagent for amide groups [42].

4.2.5 Crosslinking through ionizing radiation

Hydrogels can be chemically crosslinked using ionizing radiation. Radiation sources that emit high energy such as electron beam and gamma rays have been reported to polymerize unsaturated substances. Recently, a superabsorbent hydrogel was prepared using gamma radiation and a combination of N, N-dimethylenebisacrylamide, poly(vinylacetate-co-acrylic acid) and gelatin as crosslinking agent [43].

4.2.6 Crosslinking by free radical polymerization

Monomers with low molecular weights can be crosslinked with free radicals. However, crosslinking agents such as ethylenedimethacrylate, N, N′-methylenedi(prop-2-enamide) and melamine triacrylamide still play a role in this process [38]. Hydrogels can also be prepared with the aid of UV polymerization. This technique can be used to create a photoreversible system which allows hydrogels to degrade on exposure to UV light, thus enabling drug release [44].

4.2.7 Crosslinking using enzymes

Enzymes catalyze a number of natural crosslinking reactions resulting in stability in the extracellular matrix (ECM) of human tissue structures. The assemblage of collagen or elastin is catalyzed by lysyl oxidase which enables the formation of aldehydes from lysine present in elastin and collagen molecules. Therefore, the fabrication of hydrogels with the aid of enzyme catalysts have been under continuous study with increasing positive prospects. One of its advantages is that it has the ability to maintain substrate specificity hence reducing the possibility of unwanted side effects. Enzymes which are usually used to crosslink polymers include, tyrosinase, transglutaminase, peroxidase and sortase [45].

4.2.8 Crosslinking by grafting

Some hydrogels possess fragile physical characteristics mostly due to their method of preparation. These hydrogels may be grafted onto strong support structures to improve some of their physical properties. This is common for hydrogels prepared by bulk polymerization. Starch has been grafted with a number of vinyl monomers such as acrylic acid to increase its scope of use. This was achieved by co-polymerization of vinyl monomers with free radicals generated on the surface of starch granules [46].

5. Classes of hydrogels

Hydrogels can be classified in several ways. Some classifications are based on molecular types e.g. natural and synthetic, while others are classified on the basis of composition and function. Based on this latter classification, they can be grouped into nanogels, multifunctional hydrogels, DNA-based hydrogels, stimuli-dependent hydrogels, and hydrogels based on supramolecular systems.

5.1 Nanogels

These are polymeric nanoparticles that are made of crosslinked polymer networks with nanometer sized particulate constituents that swell upon solvent penetration. They consist of polymers of natural origin, synthetic polymers or could be derived from both classes of polymers. They are tunable systems whose characteristics like charge, size, porosity, softness and amphiphilicity can be altered by varying their chemical composition. Nanogels upon swelling, exhibit matrix properties that hold absorbed liquid, and facilitate the diffusion of entrapped solutes into a bulk exterior. They are utilized in the delivery of hydrophobic and hydrophilic drugs, as well as drugs with low and high molecular weights [47]. This is due to their amphiphilic nature in addition to other characteristics of the polymer system like temperature, density of crosslinked gels, concentration of surfactants, and the type of linkage existing in the polymer system. The presence of hydrophobic groups like hydroxyl, amide, and sulphate in their structure, helps to preserve the arrangement of the polymer system upon hydration. The synthesis of nanogels usually proceeds via polymer precursors or polymerization reactions involving monomers. One unique feature of nanogels is their ability to encapsulate more than one bioactive agent in the same carrier. This is due to their hydrophilic characteristic and excellent biocompatibility. Nanogels can be classified using several parameters like responsive behavior: stimuli-responsive and non-responsive; crosslinking: physically crosslinked and chemically crosslinked; and structure: simple, hollow, functionalized, multi-layered, hairy crosslinked, and crosslinked core shell nanogels [48].

5.2 Multifunctional hydrogels

These are composites of hydrogels and other substances like nanomaterials aimed at improving the characteristics of the hydrogel for biomedical usage. Several approaches have been employed in their production including incorporation of nanoparticles into hydrogel matrix, and the addition of a second polymer network to form an interpenetrating network, or double network. Several nanomaterials are utilized in hydrogel synthesis to enhance their mechanical properties. These include: clays, metals, metal oxides, and polymers. These combinations have found uses in different aspects of biomedicine like tissue engineering, drug delivery, wound healing, bioprinting, ocular delivery, and dental therapy [49, 50].

5.3 DNA-based hydrogels

Their pure forms are derived completely from deoxyribonucleotides. Their formation is based on interchain reactions like physical entanglement, enzymatic reactions or hydrogen bonding. Specific features of tunability, specialized responsiveness, and biodegradability make them applicable in drug delivery, biosensing, and biomedical uses. Their hydrophilicity allows the interaction with water molecules to form gel-like DNA-based materials. These DNA-based hydrogels exhibit chemical properties attributable to DNA alongside the physicochemical properties of hydrogels. DNA-based hydrogels can be functionalized with structures like i-motifs, aptamers, and G-quadruplex. DNA-based hydrogels are similar in physical nature to natural substances like chitosan, polyethylene glycols (PEGs) and extracellular matrix. This enables them to be used in several biological applications like scaffolding in tissue engineering, targeted therapy of drugs, and phase separation processes. Due to its chemical robustness, DNA-based hydrogel can be combined with other polymers like chitosan, PEG, poly(lactic-co-glycolic acid) (PLGA), acrylamide and poly(N-Isopropylacrylamide) (PNIPA) to yield hybrid DNA-based hydrogels with desirable properties. Hybrid forms of these hydrogels can be derived from crosslinking which could be physical or chemical in nature. The concentration of the crosslinker and concentration of ions in solution determine the swelling capability of DNA-based hydrogels [51].

5.4 Smart hydrogels

The capacity of hydrogels to respond to external triggers gives rise to the class of hydrogels termed smart hydrogels. These triggers can be physical, chemical or biochemical in nature. Triggers like light, temperature, pressure, electrical, and magnetic fields are physical stimuli. Factors like pH, ionic strength, and chemical agents are chemical stimulants. Biochemical/biomolecular stimuli include glucose, antigens, enzymes, and ligands [52].

5.4.1 Temperature-sensitive hydrogels

These absorb or release water according to the temperature of its environment. The degree of volume change with temperature depends on the relative proportions of their hydrophilic and hydrophobic moieties. These changes are facilitated by alterations in the crosslinking force, hydrogen bonding, and hydrophobic interactions in response to temperature changes. Thermosensitive polymers include: chitosan, agarose and hyaluronic acid which are natural polymers, and synthetic polymers like the poly(N-Isopropylacrylamide) family. They can be classified as positive or negative temperature hydrogels based on their critical solution temperature (CST) [52]. Temperature-sensitive hydrogels have been employed in wound healing [53], tissue regeneration [54], and tumor treatment [55].

5.4.2 Photosensitive hydrogels

These are attractive in biomedicine due to their ability to form hydrogels *in situ* via polymerization. Photodynamic moieties are inserted into the hydrogel structure to enable tunability. These photodynamic moieties/chromophores trap optical signals and convert them to chemical signals via photoreactions like cleavage, dimerization, and isomerization. Photosensitive hydrogels have been extensively applied in tissue engineering processes [52].

5.4.3 Pressure-responsive hydrogels

These are usually temperature or pH-sensitive. They respond to changes in the ionic concentration of a solution or its pressure to release entrapped drug molecules [56].

5.4.4 Magnetic field-sensitive hydrogels

These can be referred to as ferrogels. They possess polymer networks that are chemically crosslinked and contain fine suspensions of magnetic nanoparticles. Their volume contraction can occur at sufficient external magnetic stimuli or field strength. They have been employed in cancer diagnostics and targeted drug delivery [52].

5.5 pH-responsive hydrogels

These comprise hydrogels that contain ionizable basic or acidic groups. They are cationic when possessing amine groups, while carboxylic and sulphonic groups are present in anionic hydrogels. For anionic hydrogels, when the surrounding pH is above its pKa, deprotonation occurs. Conversely, for cationic hydrogels, when the surrounding pH is below its pKa, ionization occurs. This ionization process causes them to swell as a result of electrostatic repulsion. Examples of polymers that exhibit this behavior are chitosan, DNA, cellulose, hyaluronic acid, guar gum and poly(methacrylic acid) (PMAA). They have been applied in cell encapsulation, drug delivery, and biomedicine [52].

5.5.1 Ionic strength-sensitive hydrogels

These contain ionic monomers on a polymeric chain which respond to the ionic effects of the surrounding solution. Polyelectrolyte gels are classified in this category. Their volume changes arise from a combination of the elasticity of the hydrogel network and its ionization degree. Soluble monomers like acrylic acid and [3-(methacryloylamino) propyl] trimethylammonium chloride (MAPTEC) are used to synthesize these polymers [52]. These monomers dissociate to yield charged molecules upon immersing the hydrogels in a solution. The interaction between the ionic composition of the solution and the counterions in the hydrogel creates an electrostatic gradient that facilitates osmosis leading to swelling or shrinking of the hydrogel. Factors such as binding constant of the hydrogel monomers and their intrinsic dissociation, hydrogel polymer structure, the ionic composition of the solution, and the ionic species present influence sol-gel transitions in these hydrogels [57].

5.5.2 Biomolecule-responsive hydrogels

Glucose-responsive hydrogels are responsive to glucose concentrations in the body leading to insulin release. A novel glucose-responsive peptide hydrogel that is biocompatible and loaded with catalase, glucose oxidase, and insulin for treating diabetes has been designed [58]. Under physiologic conditions, the peptide self-assembles into a hydrogel while the formation of gluconic acid from glucose triggers insulin release from the gel matrix. This idea has also been employed in continuous glucose monitoring where an injectable fluorescent hydrogel microbead was injected under the dermis using a needle. When the concentration of glucose increases, the fluorescent intensity of the microbeads will increase [52]. Enzyme-sensitive hydrogels respond to specific enzymes which serve as biomarkers for certain disease conditions like cancer, neurodegenerative diseases, inflammatory and cardiovascular diseases [52]. For instance, hydrogels have been designed to release encapsulated drugs when exposed to high concentrations of metalloproteinases in diseased cells. Enzyme-sensitive hydrogels are capable of the instantaneous release of drugs to the target sites [59]. Antigen-responsive hydrogels utilize the antigen-antibody reaction mechanism in the body. Semi-interpenetrating network (semi-IPN) hydrogels have been formed which were impregnated with an antibody to target an antigen. Conformational changes will occur upon binding of the antigen to the antibody present on the hydrogel leading

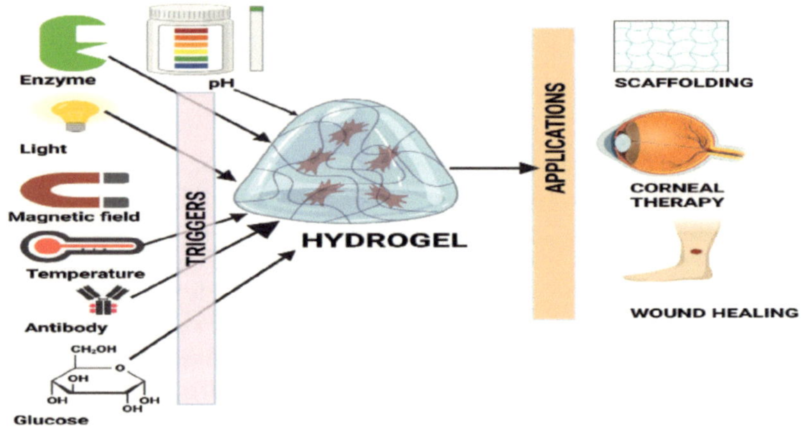

Figure 3.
Smart hydrogel system and some applications

to the release of its drug content [60]. DNA-responsive hydrogels undergo sol-gel transitions in response to the presence of DNA aptamers. Conjugating hydrogels with DNA has been employed in sensing systems and targeted drug delivery [61]. Reactive oxygen species-sensitive hydrogels are synthesized through modifications with ROS-responsive moieties either by merging them into the polymer backbone, or incorporating their side chains into the hydrogels (**Figure 3**) [62].

5.6 Supramolecular systems

The formation of supramolecular hydrogel systems involves non-covalent molecular bonds like π-π bonds, hydrogen bonding, van der Waals forces, host-guest combinations and metal-ligand binding. These supramolecular systems have enabled the development of injectable hydrogels. This is achieved by providing temporary interactions in designing polymer materials with reversible mechanical features and tunability. These supramolecular hydrogels are able to self-assemble, self-heal and respond to stimuli. In ophthalmic delivery, supramolecular hydrogels have been employed. Conventional ophthalmic delivery systems have limitations of poor adhesion and poor penetration in the aqueous humor, giving rise to sub-therapeutic concentrations in ocular tissues. Fernandes-Cunha et al. [63] designed a supramolecular hydrogel composed of adamantane in combination with hyaluronic acid or cyclodextrin for corneal wound healing. Other forms of supramolecular hydrogel systems like adhesive hydrogels, self-healing hydrogels, electrically-conducting hydrogels, and metallo-supramolecular hydrogels have been highlighted elsewhere [64].

6. Mechanism of drug entrapment and release

Drug loaded hydrogels are a group of crosslinked polymers into which one or more drugs are entrapped. During synthesis of these hydrogels, crosslinking happens either physically or chemically. The corresponding hydrogels synthesized could be termed as reversible or irreversible depending on their stability. The physical process involves weak binding interactions such as ionic bonds, hydrogen bonds and complexation. The hydrogels synthesized by this process generally have a much higher probability of deformation or degradation by changes in external parameters such as temperature, pH or physical stress while hydrogels synthesized by chemical processes are quite stable due to the strong covalent bonds holding the individual monomer units together [65]. Thus, they need to be formulated accordingly ensuring significant entrapment of the drug in the matrix. Effect of drug for a longer period of time, better bioavailability, targeting efficiency and higher contact time are some of the common advantages of hydrogel delivery systems over conventional formulations such as eye drops or other oral medications. Drugs are generally loaded in the polymer matrix either simultaneously during (in-situ encapsulation) or after crosslinking of the polymer/s. The former method encapsulates a much larger quantity of drug but there are more compatibility issues encountered due to reaction between the drug and the polymerizing monomers [65]. Drug entrapment occurs by formation of physical or chemical linkages with the polymer chains. These interactions can be classified into three types i.e. covalent, electrostatic and hydrophobic [66]. Schematic representation of these interactions are shown in **Figure 4**.

Covalent interactions are those which lead to the formation of strong chemical bonds between the drug and the polymer chain. These can either form a cleavage which on external stimulus leads to bond dissociation or remain stable which would require degradation of entire polymer matrix for release of drug [66]. Amide bonds are the most common type of highly stable covalent interactions. Ester, disulphide and peptide linkages are examples of cleavable interactions [67]. Covalent bonds are formed as a result of affinity of an electron deficient molecule towards an electron rich molecule leading to mutual sharing of electrons and bond formation. In case of amide bonds, the carboxylic group of the polymer chain is the electron deficient molecule and the amine group of the drug is the electron rich molecule. Electrostatic/ionic interactions are formed as a result of attraction of molecules having opposite charge. Unlike covalent interactions, these are non-specific and hence could be used for delivery of multiple ionic active moieties simultaneously from a single system. The intensity of these interactions controls the release rate of drug. These interactions happen by transfer of electron from one molecule to another leading to formation

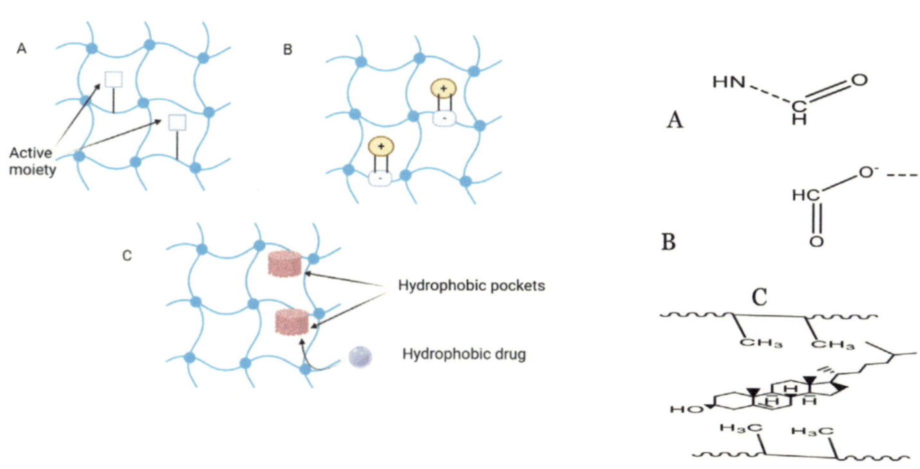

Figure 4.
Entrapment strategies (A) covalent linkage, (B) electrostatic linkage and (C) hydrophobic interactions. Linkages in the form of bonds shown by dashed lines (right).

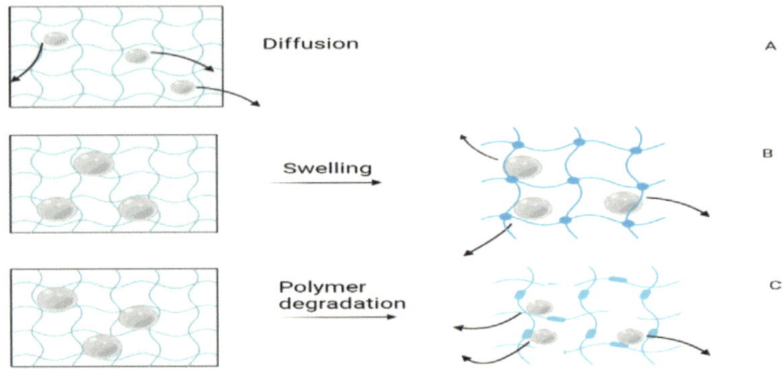

Figure 5.
Mechanisms of drug release from hydrogel matrix. (A) Simple diffusion, (B) swelling and (C) degradation.

of charge on the participating molecules. Thus, there is no electron sharing unlike covalent interactions thus, the bonds formed are much weaker. Alginate polymers containing anionic groups have been used to deliver vascular endothelial growth factor (VEGF). Sulfonate functional groups have found to improve the affinity of amino acids towards alginate polymers by improving the intensity of ionic interactions [68].

Due to the hydrophilic nature of hydrogels, incorporation of lipophilic moieties becomes very difficult. Stability and incompatibility issues such as phase separation, drug leaching during storage, degradation and erratic release profile are very common in such circumstances. Thus, there is a need for incorporation of hydrophobic functional groups into the hydrogel structures. But these hydrophobic functional groups increase the overall hydrophobicity of the hydrogel leading to hydrogel condensation thus, hampering release of drug. Cyclodextrins (CDs) are those compounds which when incorporated do not alter the hydrophilicity of hydrogels. These contain two phases- outer hydrophilic body and inner hydrophobic pocket. The active moieties bind inside the hydrophobic pocket. This facilitates efficient loading and controlled release of lipophilic moieties embedded into hydrophilic polymers [69]. Drug release rate is governed by various factors such as molecular weight of the drug, drug-polymer affinity, concentration of the polymer blend, degree of crosslinking of the polymers and the degradation rate. Three major mechanisms cater to the release of drug from the polymer. They are diffusion, swelling and degradation [66]. An appropriate polymer-polymer blend should be chosen for a particular drug of interest considering what type of release is desired. For drugs with sizes smaller than the mesh size of the hydrogel, drug release is majorly diffusion-oriented and hence would be more of an immediate release formulation but if its size exceeds the mesh size, then stearic interferences play a role in drug release. The diffusion path length also significantly increases [70] and hence drug release is sustained. In order to enable the release of such drugs through the smaller matrices, two mechanisms could be applied—swelling leading to increase in matrix size or degradation leading to breakdown of matrix network. The three drug release mechanisms are explained in **Figure 5**.

7. Biocompatibility of hydrogels for ophthalmic application

Biocompatibility of implants or delivery systems is necessary to enable appropriate activity of the substance without any local or systemic adverse events. For this, the substance needs to be compatible with the tissues to which it is going to be in contact with.

Hydrogels for ocular drug delivery should be transparent, mucoadhesive, flexible and should have the ability to retain water once administered. For administration into the aqueous chamber, the viscosity should be kept low but the viscosity should match that of vitreous fluids while administering intravitreally. The two major characteristics to consider during the development of any ocular formulation is pH and osmolality. The pH and the osmolality range for ocular drug delivery are 6.5–8.5 and 100–640 mOsm/Kg respectively [71]. Any formulation not complying with these values generally cause irritation to the eye. HET-CAM and Draize eye irritation tests are the most common tests which are carried out to evaluate the biocompatibility of ocular formulations. Both these tests give a prior idea about degree of irritation of the formulation after administration in the human eye [15]. The HET-CAM test measures the degree of hemorrhage, lysis, and coagulation on the chorioallantoic membrane of the fertilized egg of hens. The Draize test measures the degree of redness of formulation administered in the eye of rabbits, thus directly correlating to the irritation potential of the formulation. Excipients for the development of ocular formulations should be chosen in a manner that they do not cause any adverse reactions in the eye and also enable efficient activity of the API at the desired site. Choosing the appropriate excipients and also incorporating them in the right quantities in the formulation such that they cause no harm is the key. Reference to the Inactive Ingredient Guide (IIG) is done to get an idea of the upper limits of the excipients to be added into ocular formulations.

8. Modification of hydrogels for active ocular targeting

Hydrogels have found its application for both anterior and posterior segments of the eye. The anterior segment application is by soft contact lens (SCLs) for vision correction and to elute drugs, corneal wound healing, and as intraocular lenses (IOLs), whereas the posterior segment research has been as vitreous substitutes and intravitreal drug loading [72]. Vision correcting SCLs, being the most common application, generally use non-therapeutic payload on PHEMA polymer. However, evolution in the use of SCLs, hydrophilic/hydrophobic monomer hybrids of PHEMA and poly(dimethylsiloxane) to provide comfort and better structural flexibility for corneal fit and enable greater oxygen and gas permeability respectively has been achieved [73]. In an *in vitro* study by Wang et al. single protein adsorption on PEGMA—a methyl ether acrylate of poly(ethylene glycol) modified silicone hydrogel was decreased on the lens surface. This provided ophthalmic distinctness and further lowered the chances of infections [74]. Hydrogels have become a promising tool as post-surgery and ocular trauma wound closing sealant over the conventional nylon sutures. These novel ophthalmic adhesive systems have overcome the issues of ocular surface fluid leak leading to infection, anterior chamber collapse due to oozing of anterior aqueous humor and lens rotation, post IOL fitting through an incision through the cornea. ReSure® sealant is the only FDA-approved ocular healing hydrogel currently on the market, although researchers have also investigated using DuraSeal®, a formulation that is approved for cranial adhesion, to seal ocular wounds [75]. These hydrogels, which are predominantly crosslinked with PEG on trilysine amine groups, quadruple-armed with NHS capping, are shed off the eye's surface during regeneration and structurization of tissue. While the use of hydrogel based-sealants in wound healing is upsurging, researchers have been successful in loading drugs in order to enhance the healing process. Anumolu et al. designed an *in situ* forming PEGylated hydrogel loaded with doxycycline and the results ensured an accelerated healing of vesicant-induced corneal wounds [76]. Similarly, the addition of epithelial growth factors (EGF) from hydrogel formulation concluded that the released protein showed therapeutic effect in rabbit corneal ocular trauma model [77]. Hydrogel-based intraocular lenses (IOLs) have found its application in replacing the altered natural crystalline lens of the eye in cataract and even restoring the axial length of the eye in myopia. While the poly(methylmethacrylate) (PMMA) is a gold standard polymer for IOLs, the next generation silicone-based and hydrophilic/hydrophobic acrylate polymer provide greater flexibility of insertion. Other major issue with IOLs have been the posterior capsule opacification (PCO), resulting from wearing of cells from the natural lens epithelium adhering to the IOL on the posterior end. In order to avoid this issue, Bozukova et al. chemically surface-PEGylated the IOL with PEG-chains varying in molecular weights [78]. In a similar manner, Lin et al. modified the silicone hydrogel used for IOL using multilayers of hyaluronic acid and chitosan [79]. Conventionally used gases (sulphurhexafloride, perfluoropropane) and liquids (semiflourinated alkanes, silicone oils) as vitreous substitutes, though with satisfying optical and biocompatible properties, possess threats of increased intraocular pressure, choroidal thinning, inflammation, and cataracts and require a face down position for few days making it less consumer acceptable. Various hydrogel polymers concentrates have been used to replace vitreous fluid. However, their immune response, quick absorption/degradation, and inability to form tamponade, have opened the avenue for researchers to explore further modifications approaches in hydrogels as vitreous substitutes [80]. A zwitterionic *in situ* hydrogel crosslinked with α-PEGMA on a acryloyl cystamine and sulfobetaine methacrylamide copolymer, poly(MPDSA-co-AC), in a thiol-ene Michael addition reaction showed appropriate physical and rheological properties with no postoperative inflammation and remained optically acceptable *in vivo* [81].

9. Therapeutic applications of hydrogels for ocular delivery

Hydrogels have been tested for the treatment of various conditions such as macular degeneration, diabetic retinopathy, conjunctivitis, glaucoma, etc. [15, 82].

Name of product	Drug	Polymer	Application	Reference
Tiopex®	Timolol Maleate	PVA, carbomer 974P	Reduction of IOP	[15]
Pilopine HS®	Pilocarpine hydrochloride	carbopol 940	Reduction of IOP	[82]
DuraSite®	Azithromycin	poloxamer 407	Bacterial conjunctivitis	[72]
Dextenza®	Dexamethasone	Polyethylene glycol derivative	Postoperative inflammation	[82]
AktenTM®	Lidocaine hydrochloride	Hypromellose	Anesthesia	[72]
Viscotears®	—	Carbomer 980	Prevention of drying of eyes	[82]
Clinitas Gel®	—	Carbomer 980	Prevention of drying of eyes	[15]
Zirgan®	Ganciclovir	Carbomer 974P	Against herpes simplex virus	[82]

Table 1.
List of commercial hydrogel products [82].

Apart from delivery systems, hydrogels in the form of drug-eluting contact lenses, intraocular lenses, tissue adhesives as wound dressings and vitreous substitutes are alternative uses of hydrogels [72]. List of a few ocular hydrogel products in the market and their clinical uses is presented in **Table 1**.

10. Challenges and opportunities in the translation of hydrogel technology

The field of hydrogels for ocular therapy has enormously made advancement in the last few decades with multidimensional research in the field of SCLs, ocular dressings, contact lenses and smart drug delivery for ocular conditions and diseases. Though clinical translations of hydrogel technology can be seen through commercialized products for drug-free (majorly for dry-eyes syndrome) and drug-loaded hydrogel matrices, challenges are still persistent with intravitreal hydrogel-based drug delivery. **Table 2** summarizes hydrogel-based products in the market [82].

Apart from the list above, a few of them are still in various clinical trial phases before commercial approvals will be granted [82]. The difficulty of widely utilizing hydrogel inventions in the clinic for intravitreal therapy is still evident, given the challenges in the need for sterilization, scale up, cost, shelf-life, and compliance for both medical practitioners and patients. The only contender for treating AMD after intravitreal injection is OTX-TKI, a PEG hydrogel containing axitinib microcrystals—a tyrosine kinase inhibitor, with clinical studies currently in Phase 1 (NCT03630315). A sustained release variant of aflibercept called OTX-IVT, is currently being developed in the preclinical stage to bind to the circulating VEGFs and reducing the intravitreal injection frequency to once in every 4–6 months [83]. The above moieties are also being evaluated for retinal vein occlusion and diabetic macular oedema. The Japanese health authorities have recently certified ACUVUE® Theravision™ containing Ketotifen by Johnson & Johnson, the world's first drug-eluting contact lens for vision rectification and irritation of the eye due to allergic conjunctivitis for people who wear contact lenses [84].

Product	Manufacturer	Polymer	Application	Reference
Viscotear®	Novartis	Carbomer 980 (acrylic acid polymer)	Keratoconjunctivitis sicca	[82]
Lumecare®	Medicom	Carbomer 980 (acrylic acid polymer)	Keratoconjunctivitis sicca	[82]
Xailin Gel®	VISUfarma	Carbomer 980 (acrylic acid polymer)	Keratoconjunctivitis sicca	[82]
Geltears®	Bausch &Lomb	Carbomer 980 (acrylic acid polymer)	Keratoconjunctivitis sicca	[82]
ReSure® sealant	Ocular Therapeutics	PEG	Sealant post cataract surgery	[82]

Table 2.
Drug-free hydrogel-based products for ocular delivery [82].

11. Conclusion and future directions

This chapter attempted a discussion of the anatomy and physiology of the eye as an essential organ of vision. It discussed how light trapped into the ocular tissues are conveyed, relayed and processed into vision to aid physiological functions. It also exposed the various barriers which influence the circulation of molecules in the ocular architecture and what they portend for ophthalmic drug delivery. Indeed, the fabrication and use of hydrogel as an important device in the delivery of ocular therapeutics was considered in this chapter. Beyond discussions on the functionalities of smart and stimuli-responsive hydrogels presented in this chapter, it is imperative to deposit that hydrogels can be used to effect controlled drug delivery in the ocular tissue, improve topical corneal drug administration, promote prolonged corneal surface residence time, and enhance co-formulation and administration of biomacromolecules. Despite the marked successes achieved in ocular delivery using hydrogels, challenges inherent in *in vivo* delivery of biologics using hydrogels such as short half-life and poor stability, need to be tackled in order to enhance their clinical translation opportunities. Further researches are also needed to address the overwhelming influence of ocular barriers especially the anterior corneal and posterior blood retinal barriers for decreased pre-corneal clearance and improved corneal residence time, and increased transport of drug molecules into the retina. Therefore, hydrogel delivery system has vital roles to play in the future fabrication and utility of ocular therapeutics for the treatment of ophthalmic pathologies especially diseases of the posterior eye.

List of abbreviations

IOP	intraocular pressure
ABC	ATP-binding cassette
Phema	poly-(2-hydroxyethylmethacrylate)
SCLs	soft contact lenses
PVA	polyvinyl alcohol
ECM	extracellular matrix
DNA	deoxyribonucleic acid
CST	critical solution temperature
PMAA	poly(methacrylic acid)
MAPEC	3-(methacryloylamino) propyl] trimethylammonium chloride
VEGF	vascular endothelial growth factor
CD	cyclodextrins
IIG	Inactive Ingredient Guide
IOL	intraocular lenses
HET-CAM	Hen's egg chorioallantoic membrane
PEGMA	poly[oligo(ethylene glycol)methacrylate
EGF	epithelial growth factors
PMMA	poly(methylmethacrylate)
PCO	posterior capsule opacification
PEG	polyethyleneglycol

Author details

Chukwuebuka Umeyor[1,2*], Emmanuel Uronnachi[2,3], Abhit Nayak[1], Tochukwu Okeke[2], Purav Shah[4] and Vandana Patravale[1*]

1 Department of Pharmaceutical Sciences and Technology, Institute of Chemical Technology, Mumbai, India

2 Department of Pharmaceutics and Pharmaceutical Technology, Faculty of Pharmaceutical Sciences, Nnamdi Azikiwe University, Awka, Anambra State, Nigeria

3 Drug Delivery System Excellence Centre, Faculty of Pharmaceutical Sciences, Prince of Songkhla University, Thailand

4 Thoroughbred Remedies Manufacturing, TRM, Industrial Estate, Newbridge, Co Kildare, Ireland

*Address all correspondence to: ec.umeyor@unizik.edu.ng and vb.patravale@ictmumbai.edu.in

References

[1] Alvarez-Lorenzo C, Anguiano-Igea S, Varela-García A, Vivero-Lopez M, Concheiro A. Bioinspired hydrogels for drug-eluting contact lenses. Acta Biomaterialia. 2019;**84**:49-62. DOI: 10.1016/j.actbio.2018.11.020

[2] Novack GD, Robin AL. Ocular pharmacology. Journal of Clinical Pharmacology. 2016;**56**:517-527. DOI: 10.1002/jcph.634

[3] Chen X, Wu J, Lin X, Wu X, Yu X, Wang B, et al. Tacrolimus loaded cationic liposomes for dry eye treatment. Frontiers in Pharmacology. 2022;**13**:838168. DOI: 10.3389/fphar.2022.838168

[4] Muller L, Jensen BP, Bachmann LM, Wong D, Wells AP. New technique to reduce systemic side effects of timolol eye drops: The tissue press method-cross-over clinical trial. Clinical & Experimental Ophthalmology. 2020;**48**:24-30. DOI: 10.1111/ceo.13642

[5] Buwalda SJ, Boere KW, Dijkstra PJ, Feijen J, Vermonden T, Hennink WE. Hydrogels in a historical perspective: From simple networks to smart materials. Journal of Controlled Release. 2014;**190**:254-273. DOI: 10.1016/j.jconrel.2014.03.052

[6] Chang D, Park K, Famili A. Hydrogels for sustained delivery of biologics to the back of the eye. Drug Discovery Today. 2019;**24**(8):1470-1482. DOI: 10.1016/j.drudis.2019.05.037

[7] Li Q, Cao Y, Wang P. Recent advances in hydrogels for the diagnosis and treatment of dry eye disease. Gels. 2022;**8**:816. DOI: 10.3390/gels8120816

[8] Cholkar K, Dasari SR, Pal D, Mitra AK. Eye: Anatomy, physiology, and barriers to drug delivery. In: Mitra AK, editor. Ocular Transporters and Receptors. Sawston: Woodhead Publishing; 2013. pp. 1-36. DOI: 10.1533/9781908818317.1

[9] Lynch CR, Kondiah PPD, Choonara YE, du Toit LC, Ally N, Pillay V. Hydrogel biomaterials for application in ocular drug delivery. Frontiers in Bioengineering and Biotechnology. 2020;**8**:228. DOI: 10.3389/fbioe.2020.00228

[10] Gorantla S, Rapalli VK, Waghule T, Singh PP, Dubey SK, Saha RN, et al. Nanocarriers for ocular drug delivery: Current status and translational opportunity. RSC Advances. 2020;**10**:27835. DOI: 10.1039/d0ra04971a

[11] DelMonte DW, Kim T. Anatomy and physiology of the cornea. Journal of Cataract and Refractive Surgery. 2011;**37**(3):588-598. DOI: 10.1016/j.jcrs.2010.12.037

[12] Willoughby CE, Ponzin D, Ferrari S, Lobo A, Landau K, Omidi Y. Anatomy and physiology of the human eye: Effects of mucopolysaccharidoses disease on structure and function—A review. Clinical & Experimental Ophthalmology. 2010;**38**:2-11. DOI: 10.1111/j.1442-9071.2010.02363.x

[13] Harvey TM, Fernandez AGA, Patel R, Goldman D, Ciralsky J. Conjunctival anatomy and physiology. In: Holland E, Mannis M, Lee WB, editors. Ocular Surface Disease: Cornea, Conjunctiva and Tear Film. 1st ed. Philadelphia: WB Saunders; 2013. pp. 23-27

[14] Dartt DA. Regulation of mucin and fluid secretion by conjunctival

epithelial cells. Progress in Retinal and Eye Research. 2002;**21**(6):555-576. DOI: 10.1016/s1350-9462(02)00038-1

[15] Ahmed S, Amin MM, Sayed S. Ocular drug delivery: A comprehensive review. AAPS PharmSciTech. 2023;**24**(2):66. DOI: 10.1208/s12249023-02516-9

[16] Maulvi FA, Shetty KH, Desai DT, Shah DO, Willcox MDP. Recent advances in ophthalmic preparations: Ocular barriers, dosage forms and routes of administration. International Journal of Pharmaceutics. 2021;**608**:121105. DOI: 10.1016/j.ijpharm.2021.121105

[17] Danysh BP, Duncan MK. The lens capsule. Experimental Eye Research. 2009;**88**(2):151-164. DOI: 10.1016/j.exer.2008.08.002

[18] Dosmar E, Walsh J, Doyel M, Bussett K, Oladipupo A, Amer S, et al. Targeting ocular drug delivery: An examination of local anatomy and current approaches. Bioengineering. 2022;**9**(1):41. DOI: 10.3390/bioengineering9010041

[19] Norman RE, Flanagan JG, Rausch SM, Sigal IA, Tertinegg I, Eilaghi A, et al. Dimensions of the human sclera: Thickness measurement and regional changes with axial length. Experimental Eye Research. 2010;**90**(2):277-284. DOI: 10.1016/j.exer.2009.11.001

[20] Yi X, Wang Y, Yu FS. Corneal epithelial tight junctions and their response to lipopolysaccharide challenge. Investigative Ophthalmology & Visual Science;**41**(13):4093-4100

[21] Ono SJ, Abelson MB. Allergic conjunctivitis: Update on pathophysiology and prospects for future treatment. The Journal of Allergy and Clinical Immunology. 2005;**115**(1):118-122. DOI: 10.1016/j.jaci.2004.10.042

[22] Urtti A. Challenges and obstacles of ocular pharmacokinetics and drug delivery. Advanced Drug Delivery Reviews. 2006;**58**(11):1131-1135. DOI: 10.1016/j.addr.2006.07.027

[23] Mannermaa E, Vellonen KS, Urtti A. Drug transport in corneal epithelium and blood-retina barrier: Emerging role of transporters in ocular pharmacokinetics. Advanced Drug Delivery Reviews. 2006;**58**(11):1136-1163. DOI: 10.1016/j.addr.2006.07.024

[24] Kong X, Liu X, Huang X, Mao Z, Zhong Y, Chi W. Damage to the blood-aqueous barrier in eyes with primary angle closure glaucoma. Molecular Vision. 2010;**16**:2026-2032

[25] Fujiwara T, Imamura Y, Margolis R, Slakter JS, Spaide RF. Enhanced depth imaging optical coherence tomography of the choroid in highly myopic eyes. American Journal of Ophthalmology. 2009;**148**(3):445-450. DOI: 10.1016/j.ajo.2009.04.029

[26] Torres-Luna C, Fan X, Domszy R, Hu N, Wang NS, Yang A. Hydrogel-based ocular drug delivery systems for hydrophobic drugs. European Journal of Pharmaceutical Sciences. 2020;**154**:0928-0987. DOI: 10.1016/j.ejps.2020.105503

[27] Kushwaha SK, Saxena P, Rai A. Stimuli sensitive hydrogels for ophthalmic drug delivery: A review. International Journal of Pharmaceutical Investigation. 2012;**2**(2):54-60. DOI: 10.4103/2230-973X.100036

[28] Byrne ME, Park K, Peppas NA. Molecular imprinting within hydrogels. Advanced Drug Delivery Reviews. 2002;**54**(1):149-161. DOI: 10.1016/s0169-409x(01)00246-0

[29] Horne RR, Judd KE, Pitt WG. Rapid loading and prolonged release of latanoprost from a silicone hydrogel contact lens. Journal of Drug Delivery Science and Technology. 2017;**41**:410-418. DOI: 10.1016/j.jddst.2017.08.011

[30] Maulvi FA, Desai AR, Choksi HH, Patil RJ, Ranch KM, Vyas BA, et al. Effect of surfactant chain length on drug release kinetics from microemulsion-laden contact lenses. International Journal of Pharmaceutics. 2017;**524**(1-2):193-204. DOI: 10.1016/j.ijpharm.2017.03.083

[31] Xu W, Jiao W, Li S, Tao X, Mu G. Bimatoprost-loaded microemulsion laden contact lens to treat glaucoma. Journal of Drug Delivery Science and Technology. 2019;**54**:101330. DOI: 10.1016/j.jddst.2019.101330

[32] Song Y, Nagai N, Saijo S, Kaji H, Nishizawa M, Abe T. In situ formation of injectable chitosan-gelatin hydrogels through double crosslinking for sustained intraocular drug delivery. Materials Science & Engineering. C, Materials for Biological Applications. 2018;**88**:1-12. DOI: 10.1016/j.msec.2018.02.022

[33] Costa VP, Braga MEM, Duarte CMM, Alvarez-Lorenzo C, Concheiro A, Gil MH, et al. Anti-glaucoma drug-loaded contact lenses prepared using supercritical solvent impregnation. Journal of Supercritical Fluids. 2010;**53**(1-3):165-173. DOI: 10.1016/j.supflu.2010.02.007

[34] Malaekeh-Nikouei B, Ghaeni FA, Motamedshariaty VS, Mohajeri SA. Controlled release of prednisolone acetate from molecularly imprinted hydrogel contact lenses. Journal of Applied Polymer Science. 2012;**126**(1):387-394. DOI: 10.1002/app.36625

[35] Fulgencio GO, Viana FA, Ribeiro RR, Yoshida MI, Faraco AG, Cunha-Junior AS. New mucoadhesive chitosan film for ophthalmic drug delivery of timolol maleate: In vivo evaluation. Journal of Ocular Pharmacology and Therapeutics. 2012;**28**:350-358. DOI: 10.1089/jop.2011.0174

[36] Xu X, Weng Y, Xu L, Chen H. Sustained release of Avastin from polysaccharides cross-linked hydrogels for ocular drug delivery. International Journal of Biological Macromolecules. 2013;**60**:272-276. DOI: 10.1016/j.ijbiomac.2013.05.034

[37] Martens TF, Remaut K, Deschout H, Engbersen JF, Hennink WE, van Steenbergen MJ, et al. Coating nanocarriers with hyaluronic acid facilitates intravitreal drug delivery for retinal gene therapy. Journal of Controlled Release. 2015;**202**:83-92. DOI: 10.1016/j.jconrel.2015.01.030

[38] Lin J, Zheng SY, Xiao R, Yin J, Wu ZL, Zheng Q, et al. Constitutive behaviors of tough physical hydrogels with dynamic metal-coordinated bonds. Journal of the Mechanics and Physics of Solids. 2020;**139**:103935. DOI: 10.1016/j.jmps.2020.103935

[39] Yokoyama F, Masada I, Shimamura K, Ikawa T, Monobe K. Morphology and structure of highly elastic poly (vinyl alcohol) hydrogel prepared by repeated freezing-and-melting. Colloid & Polymer Science. 1986;**264**:595-601. DOI: 10.1007/BF01412597

[40] Bashir S, Hina M, Iqbal J, Rajpar AH, Mujtaba MA, Alghamdi NA, et al. Fundamental concepts of hydrogels: Synthesis, properties, and their applications. Polymers. 2020;**12**(11):2702. DOI: 10.3390/polym12112702

[41] Ali F, Khan I, Chen J, Akhtar K, Bakhsh EM, Khan SB. Emerging fabrication strategies of hydrogels and its applications. Gels. 2022;**8**:205. DOI: 10.3390/gels8040205

[42] Coviello T, Grassi M, Rambone G, Santucci E, Carafa M, Murtas E, et al. Novel hydrogel system from scleroglucan: Synthesis and characterization. Journal of Controlled Release. 1999;**60**(2-3):367-378. DOI: 10.1016/s0168-3659(99)00091-7

[43] Zu Y, Zhang Y, Zhao X, Shan C, Zu S, Wang K, et al. Preparation and characterization of chitosan-polyvinyl alcohol blend hydrogels for the controlled release of nano-insulin. International Journal of Biological Macromolecules. 2012;**50**(1):82-87. DOI: 10.1016/j.ijbiomac.2011.10.006

[44] Alla ASG, Sen M, El-Naggar AW. Swelling and mechanical properties of superabsorbent hydrogels based on Tara gum/acrylic acid synthesized by gamma radiation. Carbohydrate Polymers. 2012;**89**(2):478-485. DOI: 10.1016/j.carbpol.2012.03.031

[45] Andreopoulos FM, Beckman EJ, Russell AJ. Light-induced tailoring of PEG-hydrogel properties. Biomaterials. 1998;**9**(15):1343-1352. DOI: 10.1016/s0142-9612(97)00219-6

[46] Song W, Ko J, Choi YH, Hwang NS. Recent advancements in enzyme-mediated crosslinkable hydrogels: *In vivo*-mimicking strategies. APL Bioengineering. 2021;**5**(2):021502. DOI: 10.1063/5.0037793

[47] Ahmed EM. Hydrogel: Preparation, characterization, and applications: A review. Journal of Advanced Research. 2015;**6**(2):105-121. DOI: 10.1016/j.jare.2013.07.006

[48] Suhail M, Rosenholm JM, Minhas MU, Badshah SF, Naeem A, Khan KU, et al. Nanogels as drug-delivery systems: A comprehensive overview. Therapeutic Delivery. 2019;**10**(11):697-717. DOI: 10.4155/tde-2019-0010

[49] Barrett-Catton E, Ross ML, Asuri P. Multifunctional hydrogel nanocomposites for biomedical applications. Polymers. 2021;**13**:856. DOI: 10.3390/polym13060856

[50] Patel R, Purwada A, Cherchietti L, Inghirami G, Melnick A, Gaharwar AK, et al. Microscale bioadhesive hydrogel arrays for cell engineering applications. Cellular and Molecular Bioengineering. 2014;7:394-408. DOI: 10.1007/s12195-014-0353-8

[51] Morya V, Walia S, Mandal BB, Ghoroi C, Bhatia D. Functional DNA based hydrogels: Development, properties and biological applications. ACS Biomaterials Science & Engineering. 2020;**6**:6021-6035. DOI: 10.1021/acsbiomaterials.0c01125

[52] Merati AA, Hemmatinejad N, Shakeri M, Bashari A. Advances in research and applications of smart hydrogels part I: Preparation methods and classification. Journal of Textiles and Polymers. 2018;**6**(2):97-105

[53] Yan X, Fang WW, Xue J, Sun TC, Dong L, Zha Z, et al. Thermoresponsive in situ forming hydrogel with sol-gel irreversibility for effective methicillin-resistant *Staphylococcus aureus* infected wound healing. ACS Nano. 2019;**13**:10074-10084. DOI: 10.1021/acsnano.9b02845

[54] Zhao C, Tian S, Liu QH, Xiu KM, Lei IL, Wang Z, et al. Biodegradable nanofibrous temperature-responsive gelling microspheres for heart

regeneration. Advanced Functional Materials. 2020;**30**:2000776. DOI: 10.1002/adfm.202000776

[55] Jia YP, Shi K, Yang F, Liao JF, Han RX, Yuan LP, et al. Multifunctional nanoparticle loaded injectable thermosresponsive hydrogel as NIR controlled release platform for local photothermal immunotherapy to prevent breast cancer postoperative recurrence and metastases. Advanced Functional Materials. 2020;**30**:2001059. DOI: 10.1002/adfm.202001059

[56] Shimoboji T, Larenas E, Fowler T, Kulkarni S, Hoffman AS, Stayton PS. Photoresponsive polymer–enzyme switches. Proceedings of the National Academy of Sciences. 2002;**99**(26):16592-16596. DOI: 10.1073/pnas.262427799

[57] Li H, Lai F, Luo R. Analysis of responsive characteristics of ionic-strength-sensitive hydrogel with consideration of effect of equilibrium constant by a chemo-electro-mechanical model. Langmuir. 2009;**25**(22):13142-13150. DOI: 10.1021/la901833m

[58] Li X, Fu M, Wu J, Zhang C, Deng X, Dhinakar A, et al. pH-sensitive peptide hydrogel for glucose-responsive insulin delivery. Acta Biomaterialia. 2017;**51**:294-303. DOI: 10.1016/j.actbio.2017.01.016

[59] Zhao L, Wang J, Su D, Zhang Y, Lu H, Yan X, et al. The DNA controllable peroxidase mimetic activity of MoS2 nanosheets for constructing a robust colorimetric biosensor. Nanoscale. 2020;**12**:19420-19428. DOI: 10.1039/d0nr05649a

[60] Miyata T, Asami N, Uragami T. A reversibly antigen-responsive hydrogel. Nature. 1999;**399**(6738):766-769. DOI: 10.1038/21619

[61] Yang H, Liu H, Kang H, Tan W. Engineering target-responsive hydrogels based on aptamer–target interactions. Journal of the American Chemical Society. 2008;**130**(20):6320-6321. DOI: 10.1021/ja801339w

[62] Martin JR, Patil P, Yu F, Gupta MK, Duvall CL. Enhanced stem cell retention and antioxidative protection with injectable, ROS-degradable PEG hydrogels. Biomaterials. 2020;**263**:120377. DOI: 10.1016/j.biomaterials.2020.120377

[63] Fernandes-Cunha GM, Jeong SH, Logan CM, Le P, Mundy D, Chen F, et al. Supramolecular host-guest hyaluronic acid hydrogels enhance corneal wound healing through dynamic spatiotemporal effects. The Ocular Surface. 2022;**23**:148-161. DOI: 10.1016/j.jtos.2021.09.002

[64] Sánchez-Fernández JA. Structural strategies for supramolecular hydrogels and their applications. Polymers. 2023;**15**:1365. DOI: 10.3390/polym15061365

[65] Kirchhof S, Goepferich AM, Brandl FP. Hydrogels in ophthalmic applications. European Journal of Pharmaceutics and Biopharmaceutics. 2015;**95**:227-238. DOI: 10.1016/j.ejpb.2015.05.016

[66] Li J, Mooney DJ. Designing hydrogels for controlled drug delivery. Nature Reviews Materials. 2016;**1**:16071. DOI: 10.1038/natrevmats.2016.71

[67] Deepthi S, Jose J. Novel hydrogel-based ocular drug delivery system for the treatment of conjunctivitis. International Ophthalmology. 2019;**39**(6):1355-1366. DOI: 10.1007/s10792-018-0955-6

[68] van Hove AH, Beltejar MJ, Benoit DS. Development and in vitro assessment of enzymatically-responsive

poly(ethylene glycol) hydrogels for the delivery of therapeutic peptides. Biomaterials. 2014;**35**(36):9719-9730. DOI: 10.1016/j.biomaterials.2014.08.019

[69] Freeman I, Kedem A, Cohen S. The effect of sulfation of alginate hydrogels on the specific binding and controlled release of heparin-binding proteins. Biomaterials. 2008;**29**(22):3260-3268. DOI: 10.1016/j.biomaterials.2008.04.025

[70] Mateen R, Hoare T. Injectable, in situ gelling, cyclodextrin-dextran hydrogels for the partitioning-driven release of hydrophobic drugs. Journal of Materials Chemistry B. 2014;**2**(32):5157-5167. DOI: 10.1039/C4TB00631C

[71] Amsden B. Solute diffusion within hydrogels. Mechanisms and models. Macromolecules. 1998;**31**(23):8382-8395. DOI: 10.1021/ma980765f

[72] Cooper RC, Yang H. Hydrogel-based ocular drug delivery systems: Emerging fabrication strategies, applications, and bench-to-bedside manufacturing considerations. Journal of Controlled Release. 2019;**306**:29-39. DOI: 10.1016/j.jconrel.2019.05.034

[73] Lin HC, Yeh YH, Lin WC, Yang MC. Novel silicone hydrogel based on PDMS and PEGMA for contact lens application. Colloids and Surfaces B: Biointerfaces. 2014;**123**:986-994. DOI: 10.1016/j.colsurfb.2014.10.053

[74] Wang JJ, Liu F. Imparting antifouling properties of silicone hydrogels by grafting poly(ethylene glycol) methyl ether acrylate initiated by UV light. Journal of Applied Polymer Science. 2012;**125**(1):548-554. DOI: 10.1002/app.35687

[75] Ghobril C, Grinstaff MW. The chemistry and engineering of polymeric hydrogel adhesives for wound closure: A tutorial. Chemical Society Reviews. 2015;**44**(7):1820-1835. DOI: 10.1039/c4cs00332b

[76] Anumolu SNS, DeSantis AS, Menjoge AR, Hahn RA, Beloni JA, Gordon MK, et al. Doxycycline loaded poly(ethylene glycol) hydrogels for healing vesicant-induced ocular wounds. Biomaterials. 2010;**31**(5):964-974. DOI: 10.1016/j.biomaterials.2009.10.010

[77] Schultz CL, Morck DW. Contact lenses as a drug delivery device for epidermal growth factor in the treatment of ocular wounds. Clinical & Experimental Optometry. 2010;**93**(2):61-65. DOI: 10.1111/j.1444-0938.2010.00459.x

[78] Bozukova D, Pagnoulle C, De Pauw-Gillet MC, Desbief S, Lazzaroni R, Ruth N, et al. Improved performances of intraocular lenses by poly(ethylene glycol) chemical coatings. Biomacromolecules. 2007;**8**(8):2379-2387. DOI: 10.1021/bm0701649

[79] Lin Q, Xu X, Wang B, Shen C, Tang J, Han Y, et al. Hydrated polysaccharide multilayer as an intraocular lens surface coating for biocompatibility improvements. Journal of Materials Chemistry B. 2015;**3**(18):3695-3703. DOI: 10.1039/c5tb00111k

[80] Lin Q, Lim JYC, Xue K, Su X, Loh XJ. Polymeric hydrogels as a vitreous replacement strategy in the eye. Biomaterials. 2021;**268**:120547. DOI: 10.1016/j.biomaterials.2020.120547

[81] Chang J, Tao Y, Wang B, Bao-hua G, Hong X, Jiang Y, et al. An in situ-forming zwitterionic hydrogel as vitreous substitute. Journal of Materials Chemistry B. 2015;**3**(6):1097-1105. DOI: 10.1039/c4tb01775g

[82] Fang G, Yang X, Wang Q, Zhang A, Tang B. Hydrogels-based ophthalmic drug delivery systems for treatment of ocular diseases. Materials Science and Engineering: C. 2021;**127**:112212. DOI: 10.1016/j.msec.2021.112212

[83] Pipeline—Ocular Therapeutix. Available from: https://www.ocutx.com/research/pipeline/ [Accessed: May 14, 2023]

[84] Approval of Drug-Releasing Combination Contact Lens for Vision Correction and Allergic Eye Itch. Available from: https://www.jjvision.com/press-release/johnson-johnson-vision-receives-approval-worlds-first-and-only-drug-releasing [Accessed: May 14, 2023]

Chapter 8

Therapeutic Potentials of Hydrogels and Nanogels in CNS Disorders

Maryam Adenike Salaudeen

Abstract

Brain disorders, particularly those that worsen with age, often classified as neurodegenerative disorders constitute a major problem worldwide owing to their complexity and tremendous challenges with getting befitting therapies for them. Biomaterial technology advancements over the past few years are igniting the hope of increased success in drug discovery and development for neurological and neurodegenerative diseases. In this review, we will discuss an overview of biomaterials used in central nervous system (CNS) disorders and their contextual ideal characteristics, the use of hydrogel and nanogel biomaterials that have been explored for the treatment of various CNS disorders, and how these materials have been utilized. We shall also cover discussions on current trends associated with the use of these materials as well as challenges and prospects in this emerging field.

Keywords: hydrogels, nanogels, biomaterials, stroke, spinal cord, Alzheimer's, Parkinson, neurodegeneration

1. Introduction

Central nervous system (CNS) disorder is a term used to describe the diseases and disorders of the brain and spinal cord. Although most of these disorders share some similarities in their etiology and pathogenesis, the presentations and clinical symptoms often vary in many. The complexity of the central nervous system makes it quite difficult and almost impossible to develop drugs that can eliminate the underlying cause of most CNS disorders. Current therapies are usually only effective at reducing pain, ameliorating symptoms and improving function. Progress however has been made in certain areas, especially in CNS infectious diseases such as meningitis, and viral and bacterial encephalitis. Other CNS disorders occur particularly in the older population, albeit it is not so uncommon to find them in the younger population. This group of CNS disorders is referred to as neurodegenerative diseases (neurodegeneration in the sense that there is a progressive loss of neuronal structure and function with aging). Common neurodegenerative disorders include Alzheimer's disease (AD), Parkinson's disease (PD), Multiple Sclerosis (MS), spinal cord injury (SCI) and Amyotrophic Lateral Sclerosis (ALS). Other diseases of the brain that are not necessarily neurodegenerative but affect the neurons are termed neurological disorders. Their pathogenesis may involve the neurons directly or indirectly through other means such as cerebral blood vessels, meninges and immune cells. There are numerous Neurological disorders, common among them are stroke, Epilepsy, Schizophrenia, depression, Obsessive-compulsive disorder (OCD) and Neurodegenerative disorders.

The incidence and prevalence of most neurological disorders vary across the globe. For instance, multiple sclerosis is common in Europe, Canada and other temperate regions while diseases such as cerebral encephalitis are more predominant in Asia, Africa and South America. Others like stroke, AD, PD and epilepsy affect people worldwide, thus requiring global attention. Despite their global burden and negative impact on the world's economy, only a few drugs are available for their treatment. For instance, there are numerous anticonvulsant agents available for the management of Epilepsy, however, these drugs have drawbacks including high cost, severe side effects and poor or incomplete remission from their use. Management of stroke (the ischaemic type) is achieved with a recombinant tissue plasminogen activator (rTPA) known as alteplase. This drug has a brief time window of fewer than 5 hours and its delayed use has been associated with haemorrhagic transformation. Levodopa is the main drug, in addition to deep brain stimulation, used in the management of PD, it however causes side effects such as bradycardia and tremor and its efficacy diminishes with PD's progression. Currently, there are no drugs for treating AD – different drugs are only used to manage symptoms. With the emergence of innovative technologies and a better understanding of some novel therapies, there is hope for new drug discovery in this seemingly bleak situation.

Factors such as the complexity of the CNS, the tight regulation of substance movement across the blood–brain barrier and the presence of efflux pumps in the brain preventing therapeutic drug accumulation in the CNS, contribute to the difficulty associated with drug development for neurological diseases. Researchers are constantly innovating new formulations and technologies to circumvent these challenges. Currently, gene therapy with the aid of adeno-associated viruses, cell-based therapy and nanoformulations are being deployed as tools to manage and treat neurological disorders [1, 2]. There are currently four (4) commercially available adeno-associated virus-based gene therapy for CNS diseases [2], whereas the use of bone-marrow-derived stem cells has been employed for ages in the treatment of some types of cancers. Each of these trends led to the discovery of highly efficient drugs for CNS disorders and paved the way for renewed hope of more drug development for neurological diseases. Unfortunately, however, they are not without drawbacks, and current studies are targeted at optimizing the benefits of the new therapeutic approaches whilst eliminating or minimizing their shortfalls. For instance, despite the remarkable success of the use of AAV gene therapy for neurological diseases, this approach is associated with thrombocytopaenia, hepatic and renal toxicity and neuroinflammation of the spinal cord, following high-dose administration [2]. More so, the major routes of administration are highly invasive, making this therapeutic option less convenient and patient-friendly. Vector engineering and the use of immunosuppressive medications are now being considered as ways to bypass these obstacles. With respect to stem-cell therapy, the use of mesenchymal stem cells as a novel therapy for various diseases including neurodegenerative disorders has also met some challenges despite initial enthusiasm. Mesenchymal stem cells (MSCs) are adult stem cells that are capable of proliferation, differentiation and self-renewal. Their use dates to 1956 when leukemia was treated using a bone-marrow transplant [3]. Since then, stem cells have been used to treat other diseases like lymphoma, and autoimmune diseases like MS, restore eyesight following the repair of cornea injury [4], and COVID-19 in 2021 [5]. In contrast, the use of MSC that have proved promising in preclinical experiments for some neurodegenerative diseases has not yielded encouraging results in clinical trials. Challenges such as tumor formation, immunorejection, poor engraftment, incomplete homing and ethical constraint have been identified as factors limiting the clinical applicability of stem cells for neurological disorder [6, 7]. These factors have been attributed to the growth and study of stem cells outside their niche environment [8] and presumably explain why most promising preclinical agents do not replicate their success in clinical trials, leading to a shift in the research approach in this field. In 1999, the stroke therapy academy and industry roundtable (STAIR) put forward some recommendations to optimize the chances of preclinical agents for success in clinical trials. The need to refine preclinical experiments in terms of screening models and methods was a major highlight in the STAIR recommendation and was restated in 2009 [9]. There are various animal models of stroke, but these models often lack features to make them a robust simulation of stroke in humans. It has been proposed that preclinical stroke experiments should be designed in a manner that would enhance the clinical translation of findings. The STAIR recommendations can be extrapolated for other neurodegenerative disorders that have recorded insignificant clinical success. One important way to achieve this is to create a cellular microenvironment inside the laboratory using biomaterial technology. A cell's niche or microenvironment plays a key role in the way the cell communicates with itself (autocrine) and other cells at varying proximity to it. This microenvironment also influences mesenchymal stem cells' proliferation, differentiation, self-renewal and aging [10]. Overall, to date, there is no therapy that can halt neuronal degeneration and death, or reconstruct defective brain circuitry following a brain injury caused by trauma or diseases [11, 12].

Biomaterial technology has witnessed remarkable advancement in the last decade, making it suitable to construct, deconstruct and investigate important cellular components in cells' microenvironments. In addition to mimicking cells' microenvironment, specific biomaterial construct allows for the prediction of a drug's pharmacokinetic and pharmacodynamic profiles. This technology uses both natural and synthetic biomaterials formulated singly or in combination to improve their biological, mechanical and physicochemical properties to dub the MSC niche in the laboratory [13, 14]. Examples of natural sources of biomaterials are proteins (such as fibrin, collagen, heparin and gelatin) and polysaccharides (such as alginate, chitin, cellulose, dextran sulphate and pectin). Synthetic sources of polymeric biomaterials include polyethylene glycol (PEG), polylactic-co-glycolic acid (PLGA), polyacrylamide (PAM) [13–15]. These biomaterials have been used in different forms such as micelles, hydrogels, nanogels, nanoparticles, liposomes and dendrimers to deliver drugs to the CNS. An ideal biomaterial for medical use should be non-toxic during and after use, should be biodegradable, biocompatible, and adaptable, and should not cause inflammation [11]. Hydrogels are prepared by using ninety or more per cent (≥90%) of water, making them incredibly soft [11]. Because of their softness and very minute nano sizes, nanogels and hydrogels have been suitably used to deliver bioactive compounds such as DNA, RNA, recombinant proteins, drugs and trophic factors [16]. Further, hydrogels find use in clinical practice and experimental medicine for disease diagnosis [17], tissue repair and engineering [18], cellular mobilization [19], and as regulators of biological adhesion by serving as barrier materials [20]. Hydrogels, and by extension nanogels, find use in drug delivery because of their unique physical properties that can easily be manipulated for specific drug delivery purposes [21]. This form of drug delivery also helps to circumvent the blood–brain barrier (BBB) restriction and thus, increases the duration of drug action. Hydrogels and nanogels have also been used in delivering cells like MSCs. MSCs can be formulated in a three-dimensional hydrogel and delivered directly into the brain as pregels or administered *via* other routes for systemic effect. The former provides an avenue to accurately measure the number of cells in a particular brain region as well as promote engraftment of the delivered MSC. Additionally, the porosity of hydrogels can be tuned to allow depot drug formulation, enabling drugs to be eluted slowly [21]. Hydrogels are also biocompatible owing to their high-water content and physicochemical similarities, both mechanically and compositionally, to cell niches. Hydrogels and nanoparticles hydrogel (nanogels) are malleable to biodegradation *via* exposure internal and external stimuli (e.g., temperature, Ph, light and magnetic fields) [21] thus promoting the control of content release; stable for systemic circulation; enhanced biomedical encapsulation due internal hydrophilic nature. Because hydrogels are deformable – conforming to the shape of their applied surface, this property is exploited for targeted local drug delivery.

In this review, we shall highlight how the different properties of hydrogels are leveraged to discover and develop new diagnostic tools, drugs, and disease models for neurological disorders.

2. Ideal properties of hydrogels CNS disorders

Hydrogels have undoubtedly been employed in different capacities for an array of diseases and disorders. Regardless of the polymer source, hydrogels share some unique properties as mentioned in the introduction. However, an ideal hydrogel should possess properties that make it suitable for use for its intended application and in its target site of action. Hydrogels intended for CNS-related applications should possess specific properties as highlighted below.

2.1 Soft texture

Hydrogels are flexible water-containing matrices made from water-insoluble polymers that are chemically or physically cross-linked. Their porosity and softness make them CNS-friendly since they are unable to cause structural damage to the CNS [22]. Because their physical properties are influenced by changes in pH, osmotic pressure and temperature, hydrogels can easily be personalized for drug delivery in the CNS [23]. Thus, increasing their suitability for neurological disorders.

2.2 Blood: brain barrier permeability

A major challenge in the development of drugs for CNS diseases is the circumvention of the blood-brain barrier by the drugs [24]. Hydrogels have properties that make them similar to the brain's extracellular matrix, which enables the diffusion of drug-loaded hydrogels through the BBB, leading to effective drug delivery [25]. The deployment of hydrogels in regenerative medicine, especially for neurological disorders is owing to their ability to provide an efficient scaffold to support stem cells prior to being transplanted into the brain. Moreso, the high water content and optimum mechanical properties of hydrogels ensures that they do not cause any harm or disruption to the brain's physical structure [26].

2.3 Biocompatible and non-inflammatory

The use of stem cells as a promising regenerative tool is limited, in part by the inefficient engraftment following transplantation. Poor engraftment has been attributed to the activation of host cells' immune response [27]. Hydrogels have been shown continuously to prevent the triggering of severe host immune responses as well as act as a barrier to the activation of the host immune factors, thereby preventing rejection after stem cell transplantation [11, 28, 29]. Moreso, both the polymers and their degradation products should be compatible with the human tissue such that they do not stimulate immunogenic responses in the host. This is a very important consideration for synthetic polymers [30]. The hydrogel should also provide a support structure that favors the cells' proliferation, differentiation and survival. Biocompatibility can also be expressed as having mechanical properties similar to neural tissues [31] and architectural properties that closely mimic neural extracellular matrix [32].

2.4 Stable and biodegradable

Brain repair and regeneration is a slow process, and as such an ideal hydrogel for CNS-related function should be chemically and physically stable over an extended period [30]. This stability ensures efficient therapeutic and/or diagnostic intervention of the hydrogel. In addition to being stable, the hydrogel and its metabolite should be biodegradable i.e., easily, and completely cleared from the living system after performing their function. A non-biodegradable hydrogel polymer is likely to induce toxicity, rendering it clinically unsafe.

2.5 Scalable

Although not often considered, it is expected that the hydrogels for neurological conditions are made from polymers that can be produced in large amounts for their deployment in clinical practice [30].

3. Application of hydrogels and nanogels in neurological disease

3.1 Stroke

Formerly known as cerebrovascular accident, stroke is one of the leading causes of death and disability globally. It is caused by either occlusion of a cerebral blood vessel or a hemorrhage in the brain leading to the terms ischaemic stroke and hemorrhagic stroke respectively. Ischaemic stroke constitutes the most common of the two types and is mainly treated using thrombolytics and antiplatelets. As mentioned earlier, recombinant tissue plasminogen activator (alteplase®) is the only FDA-approved thrombolytic agent for the treatment of ischaemic stroke. Alteplase® has a short time window of between 3 and 4.5 hrs, and its use must be preceded by a CT or MRI scan for type confirmation. These limitations have led to an increase in the search for more efficacious and universally applicable drugs for ischaemic stroke. Stroke research in the context of drug discovery and development has not recorded significant success. Many promising agents in animal models of stroke have been unable to replicate their preclinical success in humans, and this unfortunate quagmire has led to refinements in the approaches to drug discovery. The use of plant extracts and herbal preparations, drug repurposing and new chemical agents among many other potential agents have been tried with no substantive success. In recent time, regenerative medicine using mesenchymal stem cells have gained some popularity as a promising approach to solving the mystery of drug discovery for stroke. Much research has been conducted using MSC from an array of sources on various animal models of stroke with hopes of clinical reciprocity. The common feature of these experiments is the employment of hydrogels and or nanogels for agent delivery, agent activation, or model creation as discussed in subsequent. Paragraphs.

MSC, along with growth factor, administered *via* the conventional intravenous or oral routes is reported in many works of literature to aid brain tissue repair and regeneration following a stroke incidence. The clinical applicability of these novel therapies is hampered by their inability to permeate the blood–brain barrier, the low survival rate of transplanted stem cells in the infarct area, inadequate homing into the site of injury and poor integration with the injured brain tissue. Tissue engineering holds a promising approach to addressing these challenges. Stroke therapy discovery is now facing some optimism as biomaterials are being used for drug delivery and model establishment. MSCs are currently formulated as hydrogels and nanoparticles to aid penetration of the BBB, increase homing in the site of injury, and ensure successful engraftment. Studies have revealed the versatility and potential benefits of hydrogels, using different biomaterials, in the management of stroke. A study by Wang and colleagues demonstrated a decrease in infarct volume and enhanced multiplication of neural progenitor stem cells (NPSC) within the MSC microenvironment when cyclosporin was delivered as a PLGA microsphere when compared with minipump-delivered cyclosporin [33]. This finding was corroborated in another study where hyaluronan/methylcellulose (HMAC) hydrogel was used to deliver erythropoietin, leading to the attenuation of tissue inflammation and migration of NPSC and mature neuroblast, and a reduction in apoptosis in mice injured cortical region [34]. Other factors such as brain-derived neurotrophic factor (BDNF) and vascular endothelial growth factor (VEGF), which play vital roles in tissue repair via angiogenesis and neurogenesis have also been delivered as extended-release formulations in various hydrogels [28, 35]. The results in primate and rodent species are positive and promising for use in humans.

Hydrogels are also used as models for drug screening because they mimic the cells' microenvironment. The 3D-bioprinting of hydrogels serves as an *in vitro* extracellular matrix for conditioning MSC towards successful tissue repair and regeneration in the in vivo cell niche. They provide an environment that aids stem cell homing, engraftment, viability, proliferation and functionality. Fibrin which is an important extracellular matrix protein was used to make a 3D scaffold that allowed the survival and regeneration of neuron-like cells derived from human endometrial stem cells. Compared to the scaffold of fibrin alone, a composite scaffold of fibrin, hyaluronic acid and laminin, enhanced biocompatibility, delayed degradation and helped maintain human NPSC viability and function [36]. Several other studies have proven the importance of growing stem cells (or cells in general) in a niche-like environment. For instance, oxidized alginate hydrogel and injectable and self-healing carboxymethyl chitosan both have elastic moduli-like brain tissue and proved to be viable 3D carriers for the transplantation of neural stem cells [37]. In another study, the hydrogel of injectable self-assembling laminin enhanced vasculature-mediated migration of immature neurons to injured brain lesions via the activation of the β1 integrin pathway [38]. Similarly, the *in vivo* injection of nerve precursor cells with Matrigel significantly reduced infarct volume, promote neuronal differentiation, and improved behavioral outcomes in rats [39]. Further, hydrogels like the protein

Figure 1.
Applications of hydrogels and nanogels in neurological disorders. Hydrogels and nanogels are employed primarily as drug-delivery molecules. However, they have been shown to be relevant in other aspects including, the diagnosis of some neurological conditions, modeling of brain disorders, CNS tissue repair, and regeneration.

hydrogel GSH (Genipin cross-linked sericin hydrogel) provide an efficient environment for the attachment and growth of neurons, whilst sericin permits axonal branching and extension together with the prevention of hypoxia-induced cell death of immature neuroblasts [40]. Thus, serving as a potential 3D carrier for tissue repair and regeneration after ischaemic injury (**Figure 1**).

3.2 Alzheimer's disease (AD)

Commonly referred to as the disease of old age, Alzheimer's disease treatment is currently limited to mainly symptomatic management using drugs from a broad class of drug families. The drugs used in the management of AD symptoms include memantine, galantamine, donepezil, tacrine, and rivastigmine. There is currently no drug to delay or reverse the progression of AD. Research into the use of hydrogels as potential therapeutic agents or aids is gaining momentum, and some experiments conducted to this effect have produced encouraging results. Several routes of hydrogel administration have been tried including intranasal, subcutaneous, and intracranial drug administrations. The use of microneedle patches has also been experimented with. Intranasal drug administration is the most utilized route of hydrogel drug administration owing to its ability to bypass the BBB, cause minimal off-target and toxic side effects, and rapidity of action.

Hydrogel of donepezil formulated by cross-linking hyaluronic acid-dopamine with PLGA with the aid of ferrous sulphate ($FeSO_4$) was designed as a sustained-release formulation for a single-dose drug administration [41]. This preparation improved the pharmacokinetic properties of donepezil after subcutaneous administration in rats. In another study, the pharmacokinetic properties particularly, the plasma concentration and area under the curve, of donepezil were significantly improved when its liposome was dispersed in chitosan hydrogel and administered intranasally in rabbits, thus buttressing the potential of hydrogels as an effective therapeutic aid for AD [42]. Timosaponin BII, an effective anti-AD agent whose use is limited by its low oral bioavailability was formulated in an in-situ hydrogel that is both ion and temperature sensitive. The formulation was shown to improve spontaneous behavior and spatial memory following intranasal administration in mice [43]. The use of tacrine, an anticholinesterase, in the management of AD is limited by several factors, most importantly its hepatotoxic side-effect. In research by Setya and colleagues in 2019, a nanoemulsion gel of tacrine was formulated, and administered via transdermal patches to rats. The nanogel had a superior pharmacokinetic and stability profile compared to conventional hydrogel and marketed capsules. The nanogel was also found to significantly improve neurobehavioral parameters in amnesic rats and did not cause hepatotoxicity [44].

Some underlying molecular causes of AD are the aggregation of amyloid-β (Aβ) proteins and oxidative stress due to an excess amount of reactive oxygen species (ROS). The search for therapeutic agents for AD is therefore sometimes tailored towards inhibiting Aβ aggregation or scavenging ROS. Some agents that have been found to inhibit this aggregation include curcumin (from *Curcuma Longa*, turmeric) and epigallocatechin-3-gallate (found in green tea extract). A nanogel formulation of these two compounds using modified hyaluronic acid caused a significant inhibition of Aβ aggregation compared to either agent alone [45]. Further, nanogel formulation of angiopep-2 modified AOC containing oxytocin was able to prevent the production of inflammatory cytokines by inhibiting microglia activation. Consequently, reducing the production of ROS [46].

3.3 Parkinson's disease (PD)

Being the second most common neurodegenerative disease, Parkinson's disease is characterized by extreme loss of dopaminergic neurons in the substantia nigra in the nigrostriatal pathway and the formation of Lewy bodies with α-synuclein. This disease faces the challenge of limited therapeutic options. Like Alzheimer's disease, PD currently has no cure, and treatment is mainly targeted at symptomatic relief using the gold standard drug levodopa. The advancement in biomaterial technology has kindled research into novel therapies for PD, with hydrogels from different sources taking the central stage. Current experiments are designed to either influence one or more of the processes involved in the pathogenesis of PD or to improve the pharmacokinetic properties of PD drugs. In this context, hydrogel and nanogels have been used as carriers to deliver drug molecules, nutrients, and various cells to the brain, employed as a scaffold for cell maturation, serve as a tool to model PD, and used for organ/tissue reconstruction. Some of the studies have also made it to the clinical trial stage.

As a means of drug delivery, thermosensitive hydrogels of poly (*N-isopropyl acrylamide*) (PNIPAM) were developed to deliver nanocrystals of magnolol, a polyphenolic compound, into the brain through the nasal cavity of mice. This preparation improved the stability of magnolol, prolonged nasal cavity resident time, and enhanced delivery to dopaminergic neurons after effectively bypassing the BBB. These advantages led to the improvement of PD symptoms in the MPTP model of PD [47]. In another study to promote the release of tyrosine, a precursor amino acid for the synthesis of dopamine, from psyllium, psyllium-containing hydrogels of acrylamide and methacrylamide were formulated. The hydrogels caused a significant improvement in the release of tyrosine from psyllium, thus serving as a potential therapy for PD [48]. Nanogel of albiflorin (a potent antioxidant and anti-inflammatory phytochemical from *Radix paeoniae Alba*) had greater stability and superior efficacy compared to the free drug, thereby offering a prospect for use in PD management [49]. Factors such as glial-derived neurotrophic factor (GDNF), which serve a beneficial purpose in the recovery from PD, have also been formulated in many different types of hydrogels as a therapy for PD in humans. For instance, in a PD human trial, a hydrogel was used for the sustained delivery of GDNF after a dopamine progenitor graft, by significantly enhancing the formation of new dopamine neurons and improving graft plasticity. Consequently, leading to functional motor improvement even after 5 months [50]. To reconstruct the nigrostriatal pathway, a collagen hydrogel housing the ventral mesencephalon and dorsal striatum was loaded with GDNF. A mild growth of tyrosine hydroxylase-positive nerve fiber in the direction of the dorsal striatum was observed after 3 weeks [51]. In a similar study by Moriarty and colleagues (2017), the intra-striatal administration of collagen hydrogel loaded with GDNF caused remarkable improvement in dopaminergic neuron survival, enhanced striatal-innervation capacity, and increased functional efficacy in rats [52].

Dopamine, the primary neurotransmitter that is deficient in PD has also been delivered to the brain in various experimental models of Parkinsonism. It's logically believed that the availability of functional dopaminergic neurons in the SN is an effective way to treat Parkinson's disease. To this effect, dopamine-containing nanogels of polyvinylpyrrolidone-polyacrylic acid caused normalization of motor activity with a significant disease-modifying effect following subchronic administration in Parkinsonian rats [53]. Similarly, dopamine-loaded nanoparticles of PLGA caused a sustained release of dopamine in the brain and significantly reversed neurobehavioral deficits after intravenous administration in Parkinsonian rats [54]. In addition to delivering dopamine, hydrogels have also been deployed for the delivery of dopamine agonists, levodopa, and ropinirole in animal models of PD to investigate the possible improvement in their pharmacokinetic and pharmacodynamic features. Ropinirole is given orally with frequent daily dosing due to poor oral bioavailability. In a study by Dudhipala and Gorre (2020), ropinirole-loaded nanogels were administered to haloperidol-induced parkinsonism rats, both orally and transdermally. The results revealed an improvement in the drug's oral and topical bioavailability, with a marked rise in the levels of catalase, glutathione, and dopamine [55]. Nanogels have been shown to improve the bioavailability of levodopa, a precursor of dopamine. Maximum recovery of levodopa was reported following the intranasal application of levodopa-loaded nanogel of chitosan [56].

The use of stem cells including mesenchymal stem cells from diverse sources to treat PD is also on the rise. The hydrogel of gelatine-PANI loaded with bone marrow-derived mesenchymal stem cells was shown to have superior efficacy when compared to BMSC alone by increasing the expression of BDNF and GDNF in the substantia nigra of mice following stereotactic injection [57]. Similarly, dopaminergic neurons derived from stem cells that were delivered in a 3D hydrogel scaffold showed greater survivability with a corresponding improved motor function in a PD model of mice compared to ordinary dopamine neurons [58].

Generally, biomaterial research in the context of drug discovery and development for Parkinson's Disease is receiving tremendous attention. Considering the stream of promising preclinical results being obtained, there is no doubt that a breakthrough is almost here.

4. Multiple sclerosis (MS)

Another disease of the CNS where biomaterial technology is finding usefulness is multiple sclerosis (MS). MS is a chronic demyelination disease of the CNS associated with a host of symptoms including loss of sensation, vision, and movements. The treatment of MS is largely achieved with anti-immune and anti-inflammatory agents. The advancement in biomaterial technology with a consequent improvement in neurological research has led to probes into the probable application of hydrogels as therapeutic, diagnostic, and disease-modeling agents for MS. Effective treatment for multiple sclerosis is yet to emerge, and stem cells hold a promising potential for use in this condition. However, as mentioned earlier, the short-lived stay of administered stem cells in the brain and BBB crossing contributes to challenges with stem cell therapy. Hydrogels have been used as a 3D scaffold to deliver differentiated stem cells to promote myelin sheath repair and remodeling in animal models of MS. The formulation of a hyaluronic acid-based hydrogel has been revealed to increase the bioavailability of bone marrow-derived mesenchymal stem cells [59]. Further, the transplantation of interleukin-10-treated dendritic cells (DCs) formulated in polyethylene glycol-based hydrogel through the neck prolonged the lifespan of DCs, prevented disease progression, and modulated the recruitment of immune cells in a preclinical model of MS [60]. The success recorded in these in vivo preclinical studies is a result of findings from previous in vitro preclinical experiments that investigated the effect of hydrogel-based three-dimensional scaffolds on stem cell proliferation and survival in their in vivo niches [61–64]. Findings from these experiments were able to demonstrate that the hydrogel culture allowed for cell adhesion, development of cell cytoskeleton, cell migration and differentiation, signal transduction, and morphogenesis of the 3D model [65].

Quite unique is the creation of biosensors for the diagnosis of MS and the detection of recovery-related problems, using hydrogels. In addition to serving as a carrier for materials that are responsive to biological signal transduction, hydrogels can in themselves, respond to stimuli thereby potentiating the signaling ability of wearable devices [66]. Biosensors designed to detect the presence of matrix metalloproteinase (MMP)-9, a vital peripheral marker of MS-associated neuroinflammation, have been designed and tried. In 2015, research led by Biela investigated the sensitivity of a hydrogel-based biosensor constructed in their lab using electrodes that were coated with enzyme-sensitive oxidized dextran that has been cross-linked with peptides. This device was able to detect MMP-9 in the presence of MMP-2, at levels between 50 and 400 ng/ml concentrations. The major limitation of the device is the delay of up to 5 minutes before biomarker detection [67]. Some years later, a similar design was constructed using poly (2-oxazoline) cross-linked with specific proteolytic peptides. This device was more able to detect MMP-9 in a lower range of between 0 and 160 nM following hydrogel degradation [68]. Currently, wearable biosensor devices are deployed to monitor balance and mobility with promising use in assessing fatigue, spasm, and tremor [66].

About a decade ago, hydrogels were employed as a tool to aid brain imaging for better diagnosis of multiple sclerosis. The tool, clear lipid-exchanged acrylamide-hybridized rigid imaging-compatible tissue-hydrogel (CLARITY), is an optical cleaning technology capable of making a 3D nanoporous hybrid of hydrogel compounds, from intact brain tissue. Thus, permitting the comprehensive visualization of the whole brain while maintaining natural fluorescence [69, 70]. All these accounts for unprecedented success in the application of hydrogels and nanogels in the diagnosis, prevention, and treatment of multiple sclerosis.

4.1 Brain tumor

Treatment of cancer is generally faced with challenges of poor penetration of anti-tumor agents, systemic/off-target side effects, and a high rate of cancer cells resistance to chemotherapeutic agents. Coupled with these is the additional obstacle of a tight physical barrier, BBB, that must be bypassed by brain-targeted chemotherapeutic agents. A handful number of malignant brain tumors exist with glioblastoma being the commonest and most deadly because of its high metastatic nature which results in recurrence even after triple therapy with resective surgery, radiotherapy, and chemotherapy. Treatment of glioblastoma is primarily achieved with the antineoplastic drug carmustine, available as injectable and implants/wafers. Carmustine administration as an implant is considered superior to the injectable form because it offers a sustained release of the active agent into the tumor cavity over a period of three weeks, and the implants protect against the degradation of carmustine [71]. Yet, wafers are not considered the ideal delivery form due to several important factors including the invasiveness of the implantation procedure, rapid release of carmustine, dislodgement of implant, local side effect of the drug, poor drug penetration, and drug resistance [71–73].

The biocompatibility and biodegradability properties of hydrogels make them a promising and effective alternative for anti-neoplastic drug delivery in cancers including brain cancers such as glioblastoma. Hydrogels have been utilized to increase drug accumulation at tumor sites leading to enhanced anti-tumor efficacy with reduced systemic and off-target toxicity. They've also been employed to enhance tumor death and immunogenicity in cancer therapy [74]. In addition to their use as drug carriers, hydrogels, and nanogels are deployed as in vitro 3D cell models for glioblastoma to aid the understanding of disease behavior as well as provide clarity on the interactions and communications within tumor microenvironment. Further, nanogels are now used, before or after surgery, to deliver anticancer drugs to the brain via intravenous injection and through the nasal cavity [75]. In line with this strategy, temperature-sensitive hydrogel loaded with epirubicin and paclitaxel nanoparticles caused remarkable tumor inhibition and lifespan prolongation when administered to mice bearing human glioma tumor. The hydrogel administration was done after surgical tumor resection [76]. Similarly, microRNA-loaded nanogels of polyglycerol scaffold were designed by Shatsberg and colleagues in 2016 for the treatment of glioblastoma in mice. The nanogels were delivered intratumorally and permitted internalization of the microRNA into the glioblastoma multiforme cells, leading to significant tumor growth inhibition and restoration of the tumor suppressor role of miR-34a in the xenograft mice [77]. In the same vein, etoposide and Olaparib-loaded bioadhesive pectin hydrogel were formulated as nanogel spray that allowed the deep penetration of the two drugs into tumor site and prevented their premature degradation [78].

To aid imaging of cancer cells in the brain, pH- and temperature-sensitive poly (w-isopropylacrylamide-co-acrylic acid) nanogels loaded with nanoparticles of citric acid-coated Fe3O4 were developed. This nanogel serves as a contrast agent for optical imaging and MRI imaging for glioma when it conjugates with Cy 5.5-labeled lactoferrin. More importantly, the contrast nanogel was devoid of any apparent toxic effect and poses no such risk in the future [79]. Research is still ongoing to develop more hydrogel-based therapies for brain tumors and the results obtained so far hold promise for efficient novel approaches.

4.2 Spinal cord injury (SCI)

Traumatic spinal cord injury results from a direct and sudden mechanical insult to the spinal cord resulting in apparent and lifelong autonomic dysfunction, paralysis, sensory impairment, and weakness [80]. Eventually, cystic cavities are formed from a progressive cascade of events including inflammation, neuronal damage, and neuronal death, leading to extracellular matrix (ECM) degeneration. Once formed, the cavities serve as barriers preventing the infiltration of beneficial cellular elements and axonal regeneration, thus constituting a major hindrance to neural regeneration after traumatic SCI. additionally, the cystic cavities also affect the transduction of electrical signals and stimulation of spinal cord tissues, further delaying or preventing axon regeneration and neural stem cell (NSC) differentiation [81, 82]. Biomaterial-based treatments have been experimented with and shown to act as bridges to narrow the cavity spaces.

In a study in 2021, an injectable and self-healing hydrogel of Fmoc-peptide and Fmoc-loaded chitosan carrying curcumin was fabricated and was shown in an in vitro test to cause Schwann cell migration away from the dorsal root ganglia, enhanced neurite growth, and caused remarkable myelination. In an in vivo test, the hydrogel promoted ECM reassembling at the site of the lesion, modified inflammatory cells to an anti-inflammatory population, and significantly improved hind limb mobility [83]. Similarly, in 2022, Luo and co-workers developed a self-healing electroconductive hydrogel ECM using chondroitin sulphate and gelatine biopolymer containing polypyrrole. The hydrogel had comparable conductive and mechanical properties with natural spinal cord tissues. In an *in vitro* experiment, the hydrogel also enhanced the outgrowth of axons, inhibited astrocyte differentiation, and promoted the differentiation of neurons. Further, the electroconductive ECM hydrogel caused a significant regeneration of myelinated axon into the lesion site, and increased neurogenesis of endogenous NSCs [84].

4.3 Epilepsy

Epilepsy is a brain disease associated with the excessive rapid firing of neurons. To alleviate the continuous electric discharges in the brain, anti-convulsant agents, like other CNS drugs, should be able to cross BBB and bind their receptors in the CNS. Conventional anticonvulsant agents are not able to efficiently circumvent this barrier, leading to higher dosage administration. Consequently, resulting in more off-target effects. The use of hydrogels and nanogels is being considered as an approach to overcome this challenge and the few preclinical experiments conducted so far have been promising. In a study by Ying and colleagues in 2014, an electro-responsive hydrogel was made from four different monomers crosslinked with *N, N′*-Methylene bisacrylamide. The hydrogel was biphasic in nature – forming a gel at high concentrations above 100 mg/ml and becoming a nanoparticle at low concentrations above 10 mg/ml but below 50 mg/ml. The hydrogel was fabricated to be brain-specific by coupling it to angiopep-2 peptide, which is a ligand of LRP (a low-density lipoprotein receptor-related protein). The specialized delivery hydrogel was used to deliver phenytoin sodium to the brains of amygdala-kindled mice, causing a significant improvement in the anticonvulsant activities of phenytoin [85].

Table 1 shows a list of hydrogels for different CNS conditions that have been approved by the US food and drug agency (FDA) or are currently under investigation in different clinical trial stages.

Currently, there is a paucity of data on the application of hydrogels and nanogels in other CNS conditions such as amyotrophic lateral sclerosis, depression, and Huntington's disease. However, the successes recorded in the use of this biomaterial approach in the discussed neurological conditions can easily be extrapolated to other CNS disorders without the need for much caution. Moreover, the challenges that were previously associated with the use of hydrogels such as uncontrolled drug release as a result of poor tunability of shape and geometry, one-time release of administered drugs and limitations with the encapsulation of some drugs such as hydroscopic drugs, antibodies, nucleic acids and proteins [21], are now being circumvented with the aid of 3D printing and the use of nanocarriers. 3D printing allows for the customization of composites of nanocarrier-hydrogel for tissue engineering and drug

Hydrogel	Polymer material	Purpose
Stroke		
GelMA [86]	Gelatine, methacrylate	Neural repair and regeneration
NeuroGel™ [87]	[*N*-2-(hydroxypropyl) methacrylamide]	Drug delivery and tissue repair
Alzheimer's disease		
Brain shuttle hydrogel [88]	Gelatine	Drug delivery
NGF Hydrogel [89, 90]	Collagen	Delay disease progression
Parkinson's Disease		
*Duodopa/Duopa [91]	Carboxymethylcellulose and polyacrylamide	Drug delivery
*BrainStem [92]	polyethylene glycol diacrylate	Drug delivery
*Levodopa/Carbidopa Gel (NeuroDerm) [93, 94]	poloxamer 188 and poloxamine 908	Drug delivery
Multiple Sclerosis		
Hyaluronic acid hydrogel [95]	Hyaluronic acid	Drug delivery and disease management
Nanoparticles hydrogel [96]	poly(lactic-co-glycolic acid) (PLGA), poly(ethylene glycol) (PEG)	Disease treatment
Brain Tumor (Glioblastoma)		
*Gliadel wafer [97–99]	poly(lactide-co-glycolide) (PLGA) and polyethylene glycol (PEG)	Disease treatment
*Proton-beam delivered polymeric hydrogel brachytherapy device [100]	polyethylene glycol (PEG), poly(lactic-co-glycolic acid) (PLGA), and Poly (2-hydroxyethyl methacrylate) (PHEMA)	Radiotherapy treatment
*Magtrace and Sentimag [101]	dextran, carboxydextran, and starch.	Disease diagnosis
Fluorescent hydrogel [102]	polyethylene glycol (PEG), polyacrylamide, hyaluronic acid, alginate, polyvinyl alcohol (PVA), and chitosan.	Tumor detection
Spinal cord injury		
NeuroScaffold [66, 103]	Poly-lactic-co-glycolic acid and Poly-l-lysine.	Supportive scaffold
Neurogel [104, 105]	Alginate	Repair and regeneration
Silk hydrogel [106]	Sericin and silk proteins derived from silkworms or spiders	Disease treatment
PuraMatrix Hydrogel [107]	RADA16 (L-arginine, L-aspartic acid, and L-alanine.)	Cell growth and tissue regeneration
Epilepsy		
Carbamazepine hydrogel [108]	Ethylcellulose, sodium alginate	Drug delivery

**Represent FDA-approved hydrogels.*

Table 1.
Some hydrogels under investigation and those approved for neurological disorders.

delivery [109]. Nanocarriers such as liposomes, micelles, dendrimers, and nanotubes that allows surface modification of drugs have also been deployed to improve drug kinetics, aiding targeted delivery [110].

5. Conclusion

Biomaterial technology is unarguably advancing at a rapid pace, and this is occurring in tandem with the increasing search for effective drugs for neurological disorders. So far, preclinical results for experiments involving hydrogels and nanogels have been encouraging and this gives a hint to the anticipation of positive clinical outcomes in a few years to come. However, a lot still needs to be done including finding ways to employ nanogels and hydrogels to improve the intranasal delivery of less potent drugs and reducing the local side effects that follow extended nasal drug delivery.

Author details

Maryam Adenike Salaudeen[1,2]

1 University of Manchester, Manchester, United Kingdom

2 Ahmadu Bello University, Zaria, Nigeria

*Address all correspondence to: masalaudeen@abu.edu.ng

References

[1] Savitz SI, Cox CS. Cell-based therapies for neurological disorders — The bioreactor hypothesis. Nature Reviews Neurology. 2023;**19**(1):9-18

[2] Kang L, Jin S, Wang J, Lv Z, Xin C, Tan C, et al. AAV vectors applied to the treatment of CNS disorders: Clinical status and challenges. Journal of Controlled Release. 2023;**355**:458-473

[3] Halme DG, Kessler DA. FDA regulation of stem-cell-based therapies. The New England Journal of Medicine. 2006;**355**(16):1730-1735

[4] Calonge M, Pérez I, Galindo S, Nieto-Miguel T, López-Paniagua M, Fernández I, Alberca M, Herreras JM. A proof-of-concept clinical trial using mesenchymal stem cells for the treatment of corneal epithelial stem cell deficiency. Translational Research. 2019;**206**:18-40

[5] Beghini DG, Horita SI, A. Henriques-pons mesenchymal stem cells in the treatment of COVID-19, a promising future. Cell. 2021;**10**(10):2588. DOI: 10.3390/cells10102588

[6] Master Z, McLeod M, Mendez I. Benefits, risks and ethical considerations in translation of stem cell research to clinical applications in Parkinson's disease. Journal of Medical Ethics. 2007;**33**(3):169-173

[7] Ikehara S. Grand challenges in stem cell treatments. Frontiers in Cell and Development Biology. 2013;**1**:2

[8] Choi JS, Harley BA. Challenges and opportunities to harnessing the (hematopoietic) stem cell niche. Current Stem Cell Report. 2016;**2**(1):85-94

[9] Fisher M, Feuerstein G, Howells DW, Hurn PD, Kent TA, Savitz SI, et al. Update of the stroke therapy academic industry roundtable preclinical recommendations. Stroke. 2009;**40**(6):2244-2250

[10] Wan PX, Wang BW, Wang ZC. Importance of the stem cell microenvironment for ophthalmological cell-based therapy. World Journal of Stem Cells. 2015;**7**(2):448-460

[11] Fernandez-Serra R, Gallego R, Lozano P, González-Nieto D. Hydrogels for neuroprotection and functional rewiring: A new era for brain engineering. Neural Regeneration Research. 2020;**15**(5):783-789

[12] Grimaudo MA, Krishnakumar GS, Giusto E, Furlani F, Bassi G, Rossi A, Molinari F, Panseri S. Panseri, Bioactive injectable hydrogels for on demand molecule/cell delivery and for tissue regeneration in the central nervous system. Acta Biomaterialia. 2022;**140**:88-101

[13] Sivashanmugam A, Arun Kumar R, Vishnu Priya M, Nair SV, Jayakumar R. An overview of injectable polymeric hydrogels for tissue engineering. European Polymer Journal. 2015;**72**:543-565

[14] Tsou YH, Khoneisser J, Huang PC, Xu X. Hydrogel as a bioactive material to regulate stem cell fate. Bioactive Materials. 2016;**1**(1):39-55

[15] Potjewyd G, Moxon S, Wang T, Domingos M, Hooper NM. Tissue engineering 3D neurovascular units: A biomaterials and bioprinting perspective. Trends in Biotechnology. 2018;**36**(4):457-472

[16] Orive G, Anitua E, Pedraz JL, Emerich DF. Biomaterials for promoting brain protection, repair and regeneration. Nature Reviews Neuroscience. 2009;**10**(9):682-692

[17] van der Linden HJ, Herber S, Olthuis W, Bergveld P. Stimulus-sensitive hydrogels and their applications in chemical (micro)analysis. Analyst. 2003;**128**(4):325-331

[18] Lee KY, Mooney DJ. Hydrogels for tissue engineering. Chemical Reviews. 2001;**101**(7):1869-1879

[19] Jen AC, Wake MC, Mikos AG. Review: Hydrogels for cell immobilization. Biotechnology and Bioengineering. 1996;**50**(4):357-364

[20] Bennett SL, Melanson DA, Torchiana DF, Wiseman DM, Sawhney AS. Next-generation hydrogel films as tissue sealants and adhesion barriers. Journal of Cardiac Surgery. 2003;**18**(6):494-499

[21] Hoare TR, Kohane DS. Hydrogels in drug delivery: Progress and challenges. Polymer. 2008;**49**(8):1993-2007

[22] Khan J, Rudrapal M, Bhat EA, Ali A, Alaidarous M, Alshehri B, Banwas S, Egbuna C. Perspective insights to bio-nanomaterials for the treatment of neurological disorders. Frontiers in Bioengineering and Biotechnology. 2021;**9**:724158

[23] Tao J, Liu H, Wu W, Zhang J, Liu S, Zhang J, Huang Y, Yang S. 3D-printed nerve conduits with live platelets for effective peripheral nerve repair. Advanced Functional Materials. 2020;**30**(42):2004272

[24] Newland B, Dunnett SB, Dowd E. Targeting delivery in Parkinson's disease. Drug Discovery Today. 2016;**21**(8):1313-1320

[25] Bordoni M, Scarian E, Rey F, Gagliardi S, Carelli S, Pansarasa O, et al. Biomaterials in neurodegenerative disorders: A promising therapeutic approach. International Journal of Molecular Sciences. 2020;**21**(9):3243

[26] Willerth SM. Bioprinting neural tissues using stem cells as a tool for screening drug targets for Alzheimer's disease. Future Medicine. 2018;**2**:163-165

[27] Cook DJ, Nguyen C, Chun HN, Llorente IL, Chiu AS, Machnicki M, Zarembinski TI, Carmichael ST. Hydrogel-delivered brain-derived neurotrophic factor promotes tissue repair and recovery after stroke. Journal of Cerebral Blood Flow & Metabolism. 2017;**37**(3):1030-1045

[28] Cook DJ, Nguyen C, Chun HN, Llorente IL, Chiu AS, Machnicki M, Zarembinski TI, Carmichael ST. Hydrogel-delivered brain-derived neurotrophic factor promotes tissue repair and recovery after stroke. Journal of Cerebral Blood Flow and Metabolism. 2017;37(3):1030-1045

[29] Nih LR, Sideris E, Carmichael ST, Segura T. Injection of microporous annealing particle (MAP) hydrogels in the stroke cavity reduces gliosis and inflammation and promotes NPC migration to the lesion. Advanced Materials. 2017;**29**(32):10

[30] Jarrin S, Cabré S, Dowd E. The potential of biomaterials for central nervous system cellular repair. Neurochemistry International. 2021;**144**:104971

[31] Zuidema JM, Rivet CJ, Gilbert RJ, Morrison FA. A protocol for rheological characterization of hydrogels for tissue engineering strategies. Journal of Biomedical Materials Research

Part B: Applied Biomaterials. 2014;**102**(5):1063-1073

[32] Niemczyk B, Sajkiewicz P, Kolbuk D. Injectable hydrogels as novel materials for central nervous system regeneration. Journal of Neural Engineering. 2018;**15**(5):051002

[33] Caicco MJ, Cooke MJ, Wang Y, Tuladhar A, Morshead CM, Shoichet MS. A hydrogel composite system for sustained epi-cortical delivery of Cyclosporin A to the brain for treatment of stroke. Journal of Controlled Release. 2013;**166**(3):197-202

[34] Wang Y, Cooke MJ, Morshead CM, Shoichet MS. Hydrogel delivery of erythropoietin to the brain for endogenous stem cell stimulation after stroke injury. Biomaterials. 2012;**33**(9):2681-2692

[35] Zhang ZG, Zhang L, Jiang Q, Zhang R, Davies K, Powers C, Bruggen N, Chopp M. VEGF enhances angiogenesis and promotes blood-brain barrier leakage in the ischemic brain. The Journal of Clinical Investigation. 2000;**106**(7):829-838

[36] Arulmoli J, Wright HJ, Phan DTT, Sheth U, Que RA, Botten GA, Keating M, Flanagan LA. Combination scaffolds of salmon fibrin, hyaluronic acid, and laminin for human neural stem cell and vascular tissue engineering. Acta Biomaterialia. 2016;**43**:122-138

[37] Wei Z, Zhao J, Chen YM, Zhang P, Zhang Q. Self-healing polysaccharide-based hydrogels as injectable carriers for neural stem cells. Scientific Reports. 2016;**6**:37841

[38] Fujioka T, Kaneko N, Ajioka I, Nakaguchi K, Omata T, Ohba H, Fässler R, Sawamoto K. Sawamoto, β1 integrin signaling promotes neuronal migration along vascular scaffolds in the post-stroke brain. eBioMedicine. 2017;**16**:195-203

[39] Jin K, Mao X, Xie L, Galvan V, Lai B, Wang Y, Gorostiza O, Greenberg DA. Transplantation of human neural precursor cells in Matrigel scaffolding improves outcome from focal cerebral ischemia after delayed postischemic treatment in rats. Journal of Cerebral Blood Flow and Metabolism. 2010;**30**(3):534-544

[40] Wang Z, Wang J, Jin Y, Luo Z, Yang W, Xie H, Huang K, Wang L. A neuroprotective Sericin hydrogel As an effective neuronal cell carrier for the repair of ischemic stroke. ACS Applied Materials & Interfaces. 2015;**7**(44):24629-24640

[41] Lee SY, Park J-H, Yang M, Baek M.-J, Kim M-H, Lee J, Khademhosseini A, Cho H-J. Ferrous sulfate-directed dual-cross-linked hyaluronic acid hydrogels with long-term delivery of donepezil. International Journal of Pharmaceutics. 2020;**582**:119309

[42] Al Harthi S, Alavi SE, Radwan MA, El Khatib MM, AlSarra IA. Nasal delivery of donepezil HCl-loaded hydrogels for the treatment of Alzheimer's disease. Scientific Reports. 2019;**9**(1):9563

[43] Chen W, Li R, Zhu S, Ma J, Pang L, Ma B, Du L, Jin Y. Nasal timosaponin BII dually sensitive in situ hydrogels for the prevention of Alzheimer's disease induced by lipopolysaccharides. International Journal of Pharmaceutics. 2020;**578**:119115

[44] Setya S, Madaan T, Razdan BK, Farswan M, Talegaonkar S. Design and development of novel transdermal Nanoemulgel for Alzheimer's disease: Pharmacokinetic, Pharmacodynamic and

biochemical investigations. Current Drug Delivery. 2019;**16**(10):902-912

[45] Zhang Y, Yang H, Wei D, Zhang X, Wang J, Wu X, et al. Mitochondria-targeted nanoparticles in treatment of neurodegenerative diseases. Exploration. 2021;**1**(3):20210115

[46] Zhang Y, Zou Z, Liu S, Miao S, Liu H. Nanogels as novel Nanocarrier Systems for Efficient Delivery of CNS therapeutics. Frontiers in Bioengineering and Biotechnology. 2022;**10**:954470

[47] TanY, Liu Y, Liu Y, Ma R, Luo J, Hong H, Chen X, Chen T. Rational Design of Thermosensitive Hydrogel to deliver nanocrystals with intranasal Administration for Brain Targeting in Parkinson's disease. Research (Wash D C) 2021;**2021**:9812523

[48] Singh B, Chauhan N. Release dynamics of tyrosine from dietary fiber psyllium based hydrogels for use in Parkinson's disease. Food Research International. 2010;**43**(4):1065-1072

[49] Chen Y-B, Qiao T, Wang Y-Q, Cui Y-L, Wang Q-S. Hydrogen bond-enhanced nanogel delivery system for potential intranasal therapy of Parkinson's disease. Materials & Design. 2022;**219**:110741

[50] Hunt CPJ, Penna V, Gantner CW, Moriarty N, Wang Y, Franks S, Ermine CM, Parish CL. Tissue programmed hydrogels functionalized with GDNF improve human neural grafts in Parkinson's disease. Advanced Functional Materials. 2021;**31**(47):2105301

[51] Ucar B, Kajtez J, Foidl BM, Eigel D, Werner C, Long KR, Emnéus J, Humpel C. Biomaterial based strategies to reconstruct the nigrostriatal pathway in organotypic slice co-cultures. Acta Biomaterialia. 2021;**121**:250-262

[52] Moriarty N, Pandit A, Dowd E. Encapsulation of primary dopaminergic neurons in a GDNF-loaded collagen hydrogel increases their survival, re-innervation and function after intra-striatal transplantation. Scientific Reports. 2017;7(1):1-14

[53] Rashed ER, Abd El-Rehim HA, El-Ghazaly MA. Potential efficacy of dopamine loaded-PVP/PAA nanogel in experimental models of parkinsonism: Possible disease modifying activity. Journal of Biomedical Materials Research Part A. 2015;**103**(5):1713-1720

[54] Pahuja R, Seth K, Shukla A, Shukla RK, Bhatnagar P, Chauhan LKS, Saxena PN, Gupta KC. Trans-blood brain barrier delivery of dopamine-loaded nanoparticles reverses functional deficits in parkinsonian rats. ACS Nano. 2015;**9**(5):4850-4871

[55] Dudhipala N, Gorre T. Neuroprotective effect of Ropinirole lipid nanoparticles enriched hydrogel for Parkinson's disease: In vitro, ex vivo, pharmacokinetic and Pharmacodynamic evaluation. Pharmaceutics. 2020;**12**(5):448

[56] Sharma S, Lohan S, Murthy RS. Formulation and characterization of intranasal mucoadhesive nanoparticulates and thermo-reversible gel of levodopa for brain delivery. Drug Development and Industrial Pharmacy. 2014;**40**(7):869-878

[57] Xue J, Liu Y, Darabi MA, Tu G, Huang L, Ying L, Xiao B, Zhang L. An injectable conductive gelatin-PANI hydrogel system serves as a promising carrier to deliver BMSCs for Parkinson's disease treatment. Materials Science and Engineering: C. 2019;**100**:584-597

[58] Francis NL, Zhao N, Calvelli HR, Saini A, Gifford JJ, Wagner GC, Cohen RI,

Moghe PV. Peptide-based scaffolds for the culture and transplantation of human dopaminergic neurons. Tissue Engineering. Part A. 2020;**26**(3-4):193-205

[59] Ferreira H, Amorim D, Lima AC, Pirraco RP, Costa-Pinto AR, Almeida R, Almeida A, Neves NM. A biocompatible and injectable hydrogel to boost the efficacy of stem cells in neurodegenerative diseases treatment. Life Sciences. 2021;**287**:120108

[60] Thomas AM, Beskid NM, Blanchfield JL, Rosado AM, García AJ, Evavold BD, et al. Localized hydrogel delivery of dendritic cells for attenuation of multiple sclerosis in a murine model. Journal of Biomedical Materials Research. Part A. 2021;**109**(7):1247-1255

[61] Jahanbazi Jahan-Abad A, Karima S, Sahab Negah S, Noorbakhsh F, Borhani-Haghighi M, Gorji A. Therapeutic potential of conditioned medium derived from oligodendrocytes cultured in a self-assembling peptide nanoscaffold in experimental autoimmune encephalomyelitis. Brain Research. 2019;**1711**:226-235

[62] Unal DB, Caliari SR, Lampe KJ. 3D hyaluronic acid hydrogels for modeling oligodendrocyte progenitor cell behavior as a function of matrix stiffness. Biomacromolecules. 2020;**21**(12):4962-4971

[63] Zambutot SG, Serranot JF, Vilbert AC, Lu Y, Harley BAC, Pedron S. Response of neuroglia to hypoxia-induced oxidative stress using enzymatically crosslinked hydrogels. MRS Communications. 2020;**10**(1):83-90

[64] Baisiwala S, Moreno MA, Wang C, Rogan HAW, Tsai H-C, Yang F, et al. A 3-dimensional hydrogel model of multiple sclerosis brain lesions reveals insights into Re-myelination. Journal of Biomaterials and Tissue Engineering. 2017;**7**(9):868-875

[65] Schindler M, Nur-E-Kamal A, Ahmed I, Kamal J, Liu H-Y, Amor N, Ponery AS, Meiners S. Meiners, Living in three dimensions. Cell Biochemistry and Biophysics 2006;**45**(2):215-227

[66] Liu X, Ren Z, Liu F, Zhao L, Ling Q, Gu H. Multifunctional self-healing dual network hydrogels constructed via host-guest interaction and dynamic covalent bond as wearable strain sensors for monitoring human and organ motions. ACS Applied Materials & Interfaces. 2021;**13**(12):14612-14622

[67] Biela A, Watkinson M, Meier UC, Baker D, Giovannoni G, Becer CR, et al. Disposable MMP-9 sensor based on the degradation of peptide cross-linked hydrogel films using electrochemical impedance spectroscopy. Biosensors and Bioelectronics. 2015;**68**:660-667

[68] Ahmad N, Colak B, Gibbs MJ, Zhang D-W, Gautrot JE, Watkinson M, Becer CR, Krause S. Peptide cross-linked poly(2-oxazoline) as a sensor material for the detection of proteases with a quartz crystal microbalance. Biomacromolecules. 2019;**20**(7):2506-2514

[69] Chung K, Wallace J, Kim S-Y, Kalyanasundaram S, Andalman AS, Davidson TJ, Mirzabekov JJ, Deisseroth K. Deisseroth, Structural and molecular interrogation of intact biological systems. Nature. 2013;**497**(7449):332-337

[70] Tomer R, Ye L, Hsueh B, Deisseroth K. Advanced CLARITY for rapid and high-resolution imaging of intact tissues. Nature Protocols. 2014;**9**(7):1682-1697

[71] Juratli TA, Schackert G, Krex D. Current status of local therapy in malignant gliomas--a clinical review of three selected approaches. Pharmacology & Therapeutics. 2013;**139**(3):341-358

[72] Wolinsky JB, Colson YL, Grinstaff MW. Local drug delivery strategies for cancer treatment: Gels, nanoparticles, polymeric films, rods, and wafers. Journal of Controlled Release. 2012;**159**(1):14-26

[73] Perry J, Chambers A, Spithoff K, Laperriere N. Gliadel wafers in the treatment of malignant glioma: A systematic review. Current Oncology. 2007;**14**(5):189-194

[74] Sun M, Yue T, Wang C, Fan Z, Gazit E, Du J. Ultrasound-responsive peptide Nanogels to balance conflicting requirements for deep tumor penetration and prolonged blood circulation. ACS Nano. 2022;**16**(6):9183-9194

[75] Stawicki B, Schacher T, Cho H. Nanogels as a versatile drug delivery system for brain cancer. Gels. 2021;7:63. DOI: 10.3390/gels7020063

[76] Lin FW, Chen PY, Wei KC, Huang CY, Wang CK, Yang HW. Rapid In situ MRI traceable gel-forming dual-drug delivery for synergistic therapy of brain tumor. Theranostics. 2017;7(9):2524-2536

[77] Shatsberg Z, Zhang X, Ofek P, Malhotra S, Krivitsky A, Scomparin A, Tiram G, Satchi-Fainaro R. Functionalized nanogels carrying an anticancer microRNA for glioblastoma therapy. Journal of Controlled Release. 2016;**239**:159-168

[78] McCrorie P, Mistry J, Taresco V, Lovato T, Fay M, Ward I, Ritchie AA. Etoposide and olaparib polymer-coated nanoparticles within a bioadhesive sprayable hydrogel for post-surgical localised delivery to brain tumours. European Journal of Pharmaceutics and Biopharmaceutics. 2020;**157**:108-120

[79] Jiang L, Zhou Q, Mu K, Xie H, Zhu Y, Zhu W, Zhao Y, Yang X. pH/temperature sensitive magnetic nanogels conjugated with Cy5.5-labled lactoferrin for MR and fluorescence imaging of glioma in rats. Biomaterials. 2013;**34**(30):7418-7428

[80] Dell'Anno MT, Wang X, Onorati M, Li M, Talpo F, Sekine Y, Ma S, Strittmatter SM. Strittmatter, Human neuroepithelial stem cell regional specificity enables spinal cord repair through a relay circuit. Nature Communications. 2020;**12**(15):17207-17219

[81] Liu H, Xu X, Tu Y, Chen K, Song L, Zhai J, Chen S, He L. Engineering microenvironment for endogenous neural regeneration after spinal cord injury by reassembling extracellular matrix. ACS Applied Materials & Interfaces. 2020;**12**(15):17207-17219

[82] David G, Mohammadi S, Martin AR, Cohen-Adad J, Weiskopf N, Thompson A, et al. Traumatic and nontraumatic spinal cord injury: Pathological insights from neuroimaging. Nature Reviews. Neurology. 2019;**15**(12):718-731

[83] Luo J, Shi X, Li L, Tan Z, Feng F, Li J, Pang M, He L. An injectable and self-healing hydrogel with controlled release of curcumin to repair spinal cord injury. Bioactive Materials. 2021;6(12):4816-4829

[84] Luo Y, Fan L, Liu C, Wen H, Wang S, Guan P, Chen D, Tan G. An injectable, self-healing, electroconductive extracellular matrix-based hydrogel for enhancing tissue repair after traumatic spinal cord injury. Bioactive Materials. 2022;**7**:98-111

[85] Ying X, Wang Y, Liang J, Yue J, Xu C, Lu L, Xu Z, Chen Z. Angiopep-conjugated electro-responsive hydrogel nanoparticles: Therapeutic potential for epilepsy. Angewandte Chemie (International Ed. in English). 2014;**53**(46):12436-12440

[86] Liu Y, Long L, Zhang F, Hu X, Zhang J, Hu C, Wang Y, Xu J. Microneedle-mediated vascular endothelial growth factor delivery promotes angiogenesis and functional recovery after stroke. Journal of Controlled Release. 2021;**338**:610-622

[87] Ma X, Wang M, Ran Y, Wu Y, Wang J, Gao F, Liu Z, Feng Z. Design and fabrication of polymeric hydrogel carrier for nerve repair. Polymers. 2022;**14**(8):1549

[88] Weber F, Bohrmann B, Niewoehner J, Fischer JAA, Rueger P, Tiefenthaler G, Moelleken J, Freskgård PO. Brain shuttle antibody for Alzheimer's disease with attenuated peripheral effector function due to an inverted binding mode. Cell Reports. 2018;**22**(1):149-162

[89] Foidl BM, Ucar B, Schwarz A, Rebelo AL, Pandit A, Humpel C. Nerve growth factor released from collagen scaffolds protects axotomized cholinergic neurons of the basal nucleus of Meynert in organotypic brain slices. Journal of Neuroscience Methods. 2018;**295**:77-86

[90] Humpel C. NGF released from blood cells or collagen hydrogels as a therapeutic target in Alzheimer's disease? Advances in Experimental Medicine and Biology. 2021;**1331**:193-202

[91] Olanow, C.W., K. Kieburtz, P. Odin, A.J. Espay, D.G. Standaert, H.H. Fernandez, A. Vanagunas, . . . A. Antonini, Continuous intrajejunal infusion of levodopa-carbidopa intestinal gel for patients with advanced Parkinson's disease: A randomised, controlled, double-blind, double-dummy study. Lancet Neurology, 2014. 13(2): p. 141-149

[92] Paccione N, Rahmani M, Barcia E, Negro S. Antiparkinsonian agents in investigational polymeric micro- and Nano-Systems. Pharmaceutics. 2022;**15**(1):13

[93] Olanow, C.W., A.J. Espay, F. Stocchi, A.L. Ellenbogen, M. Leinonen, L. Adar, R.J. Case, . . . W. Poewe, Continuous subcutaneous levodopa delivery for Parkinson's disease: A randomized study. Journal of Parkinson's Disease, 2021. 11(1): p. 177-186

[94] ClinicalTrials.gov, Efficacy, Safety and Tolerability Study of ND0612 vs. Oral IR-LD/CD in Subjects With PD Experiencing Motor Fluctuations (BouNDless), N.U.S.N.L.o. Medicine, Editor. 2023

[95] Grieco M, Ursini O, Palamà IE, Gigli G, Moroni L, Cortese B. HYDRHA: Hydrogels of hyaluronic acid. New biomedical approaches in cancer, neurodegenerative diseases, and tissue engineering. Materials Today Bio. 2022;**17**:100453

[96] Lima AF, Amado IR, Pires LR. Poly(d,l-lactide-co-glycolide) (PLGA) nanoparticles loaded with proteolipid protein (PLP)-exploring a new administration route. Polymers (Basel). 2020;**12**(12):3063

[97] Panigrahi M, Das PK, Parikh PM. Brain tumor and Gliadel wafer treatment. Indian Journal of Cancer. 2011;**48**(1):11-17

[98] Abel TJ, Ryken T, Lesniak MS, Gabikian P. Gliadel for brain metastasis. Surgical Neurology International. 2013;**4**(Suppl. 4):S289-S293

[99] Iuchi T, Inoue A, Hirose Y, Morioka M, Horiguchi K, Natsume A, Arakawa Y, Sakata Y. Long-term effectiveness of Gliadel implant for malignant glioma and prognostic factors for survival: 3-year results of a postmarketing surveillance in Japan. Neuro-Oncology Advances. 2022;4(1):189

[100] Dong YC, Bouché M, Uman S, Burdick JA, Cormode DP. Detecting and monitoring hydrogels with medical imaging. ACS Biomaterials Science & Engineering. 2021;7(9):4027-4047

[101] Administration, U.F.a.D., Magtrace and Sentimag Magnetic Localization System - P160053/S002, in Recently approved devices. 2022

[102] Hama R, Ulziibayar A, Reinhardt JW, Watanabe T, Kelly J, Shinoka T. Recent developments in biopolymer-based hydrogels for tissue engineering applications. Biomolecules. 2023;**13**(2):280

[103] Slotkin JR, Pritchard CD, Luque B, Ye J, Layer RT, Lawrence MS, O'Shea TM, Langer R. Biodegradable scaffolds promote tissue remodeling and functional improvement in non-human primates with acute spinal cord injury. Biomaterials. 2017;**123**:63-76

[104] Ansorena E, De Berdt P, Ucakar B, Simón-Yarza T, Jacobs D, Schakman O, Jankovski A, Préat V. Injectable alginate hydrogel loaded with GDNF promotes functional recovery in a hemisection model of spinal cord injury. International Journal of Pharmaceutics. 2013;**455**(1-2):148-158

[105] Ashton RS, Banerjee A, Punyani S, Schaffer DV, Kane RS. Scaffolds based on degradable alginate hydrogels and poly (lactide-co-glycolide) microspheres for stem cell culture. Biomaterials. 2007;**28**(36):5518-5525

[106] Liu Y, Zhang Z, Zhang Y, Luo B, Liu X, Cao Y, et al. Construction of adhesive and bioactive silk fibroin hydrogel for treatment of spinal cord injury. Acta Biomaterialia. 2023;**158**:178-189

[107] Gelain F, Luo Z, Rioult M, Zhang S. Self-assembling peptide scaffolds in the clinic. Npj. Regenerative Medicine. 2021;**6**(1):9

[108] Mawazi SM, Al-Mahmood SMA, Chatterjee B, Hadi HA, Doolaanea AA. Carbamazepine gel formulation as a sustained release epilepsy medication for pediatric use. Pharmaceutics. 2019;**11**(10):488

[109] Chansoria P, Etter EL, Nguyen J. Regenerating dynamic organs using biomimetic patches. Trends in Biotechnology. 2022;**40**(3):338-353

[110] Din FU, Aman W, Ullah I, Qureshi OS, Mustapha O, Shafique S, et al. Effective use of nanocarriers as drug delivery systems for the treatment of selected tumors. International Journal of Nanomedicine. 2017;**12**:7291-7309